BLUE HOPE

BLUE HOPE

EXPLORING AND CARING
FOR EARTH'S MAGNIFICENT OCEAN

SYLVIA A. EARLE

NATIONAL GEOGRAPHIC EXPLORER-IN-RESIDENCE
& FOUNDER OF MISSION BLUE

NATIONAL GEOGRAPHIC

WASHINGTON, D.C.

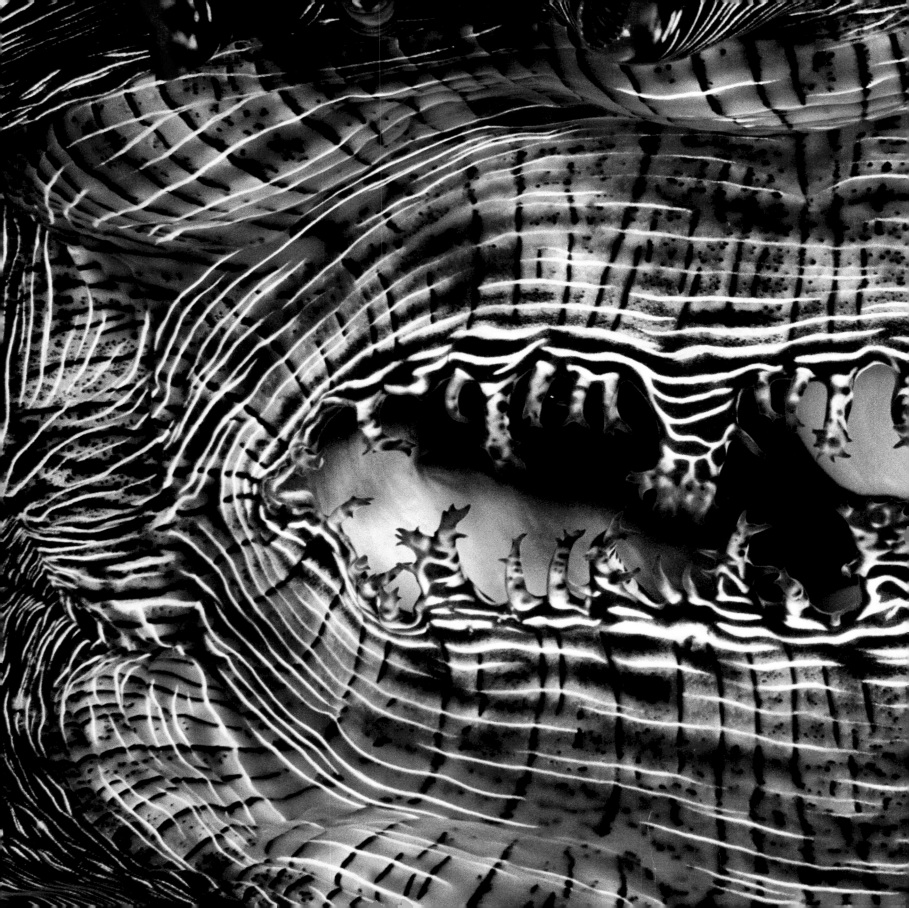

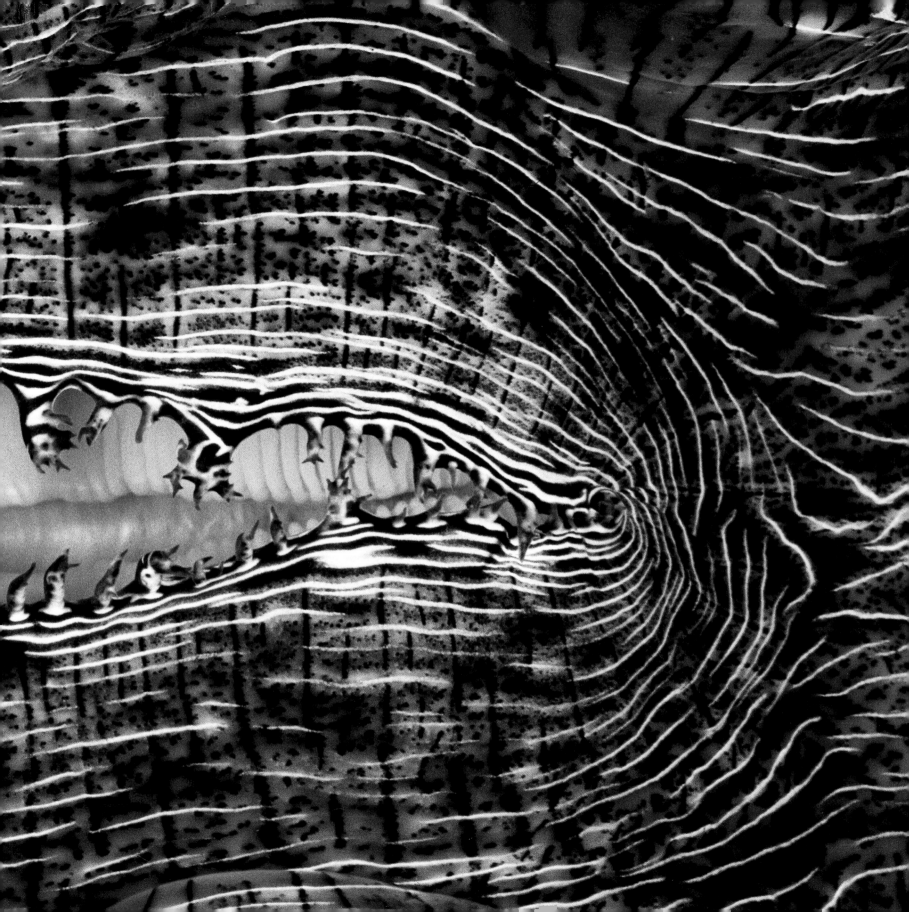

CONTENTS

Page 2: A manta ray engages divers at Flower Garden Banks National Marine Sanctuary off the Texas coast. Pages 4–5: A natural palette of multicolored coral graces the protected waters of Kingman Reef, part of the U.S. Line Islands in the North Pacific Ocean. Pages 6–7: Grazing yellowfin surgeonfish move en masse along Nikumaroro Island in the western Pacific Ocean. Pages 8–9: Warm within their feathered jackets, chinstrap penguins crown an iceberg near subantarctic Candlemas Island in the South Atlantic Sandwich Islands. Pages 10–11: In Lembeh Strait, Indonesia, microscopic algae color the opalescent lips of a giant clam. Pages 12–13: A mother humpback whale and her calf swim in a safe haven, the Hawaiian Islands Humpback Whale National Marine Sanctuary, before journeying thousands of perilous miles to Alaskan feeding areas.

INTRODUCTION
A WISH BIG ENOUGH TO CHANGE THE WORLD

If someone asked you to make a wish "big enough to change the world" and offered to help make it come true, what would *you* say? Such a call came to me from Chris Anderson, curator of the phenomenon called TED—short for Technology, Entertainment, Design, an organization that grew from a small group that gathered to discuss timely issues of global importance to an international forum with speakers making the "talk of their lives" to promote "ideas worth spreading." I was asked to give such a presentation in 18 minutes on a vast stage at the Convention Center in Long Beach, California, in February 2009—concluding with my wish.

In a perfect storm of opportunity, the TED conference converged on the same week that culminated three years of effort to help launch a new way of looking at the world via the highly acclaimed virtual globe, map, and geographical information system, Google Earth. During a conference in Spain, I had unintentionally provoked Google Earth's leader, John Hanke, to consider what that remarkable system lacked. After publicly praising the wondrous ability to "hold the world in your hands" and vicariously fly from high in the sky to your backyard, your neighbor's backyard, to the Grand Canyon, and beyond, I suggested that something big was missing—like most of the world. You should call it "Google Dirt," I said, then realized that I had likely just insulted someone I deeply admired for all that he and his team had done to enable people to see the world—and themselves—with new eyes.

But rather than take offense, Hanke invited me to speak at the Googleplex, meet the "Googlers," organize an international team of experts, and help bring about a revolution in the quantity and quality of ocean images and data available to the public via Google Ocean. My daughter, Elizabeth Taylor, and son-in-law, Ian Griffith,

were enlisted to manage the endeavor through their company, Deep Ocean Exploration and Research (DOER Marine Operations), including negotiating arrangements with the U.S. Navy to obtain the best publicly available bathymetry. Images developed through the collaboration with Google and the National Geographic Society provided a wondrous background for my TED talk and wish:

> *I wish you would use all means at your disposal—Films! Expeditions! The web! New submarines!—to create a campaign to ignite public support for a global network of marine protected areas, Hope Spots large enough to save and restore the ocean, the blue heart of the planet.*

I added, "My wish is a big wish, but if we can make it happen, it truly can change the world and help ensure the survival of my favorite species: human beings. For the children of today, for tomorrow's child . . . now is the time."

This volume describes the wish-in-progress, the development of efforts under way to take care of the living ocean that makes our existence possible. It is a standing invitation to all to participate by identifying and protecting places that are pristine, as well as those that are distressed but that, with care, can recover. Early in the 20th century, an idea "big enough to change the world" led to establishing national parks, protected areas, and policies that safeguard land, air, water, and wildlife, with results that are paying off handsomely: a healthier, more productive, prosperous, safer, and aesthetically more pleasing world.

But something big has been missing: the ocean.

BRIAN SKERRY ~ HONSHU, JAPAN, NORTH PACIFIC OCEAN
Trash to some, but boon for a yellow goby, a corroded soda can provides shelter at Izu Peninsula on Honshu island.

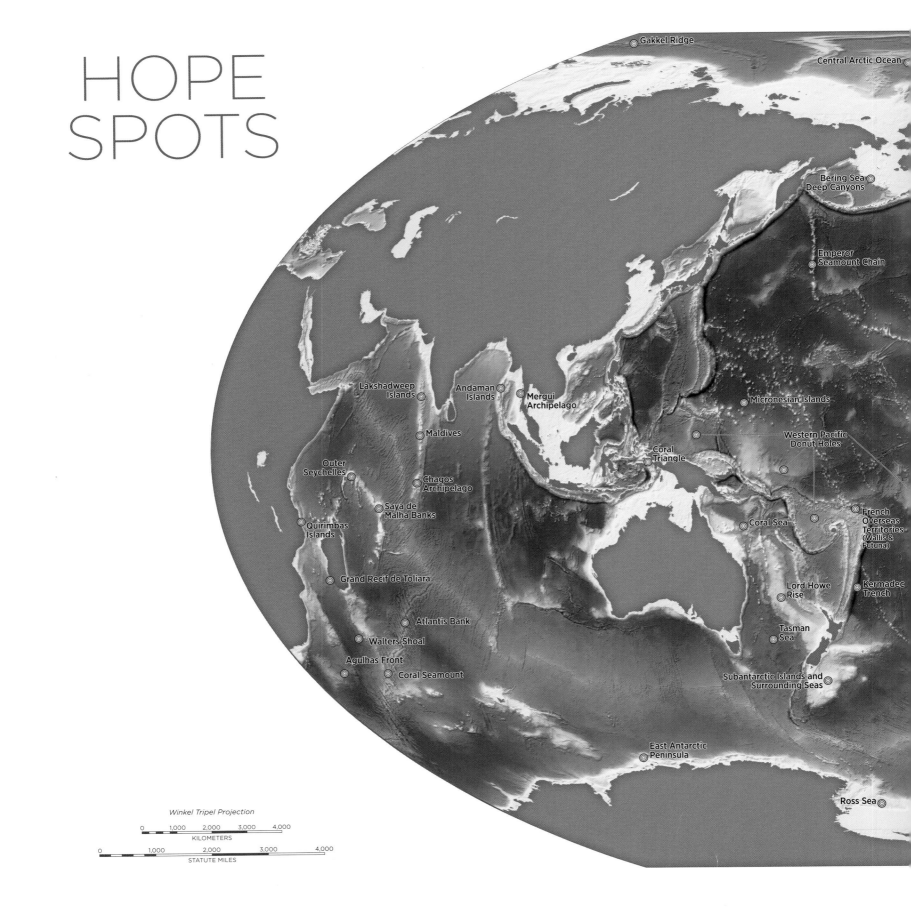

HOPE
SPOTS

Gakkel Ridge

Central Arctic Ocean

Bering Sea
Deep Canyons

Emperor
Seamount Chain

Lakshadweep
Islands

Andaman
Islands

Mergui
Archipelago

Micronesian Islands

Maldives

Western Pacific
Donut Holes

Coral
Triangle

Outer
Seychelles

Chagos
Archipelago

Saya de
Malha Banks

French
Overseas
Territories
(Wallis &
Futuna)

Coral Sea

Quirimbas
Islands

Grand Recif de Toliara

Lord Howe
Rise

Kermadec
Trench

Atlantis Bank

Tasman
Sea

Walters Shoal

Agulhas Front

Coral Seamount

Subantarctic Islands and
Surrounding Seas

East Antarctic
Peninsula

Ross Sea

Winkel Tripel Projection

0   1,000   2,000   3,000   4,000
KILOMETERS

0   1,000   2,000   3,000   4,000
STATUTE MILES

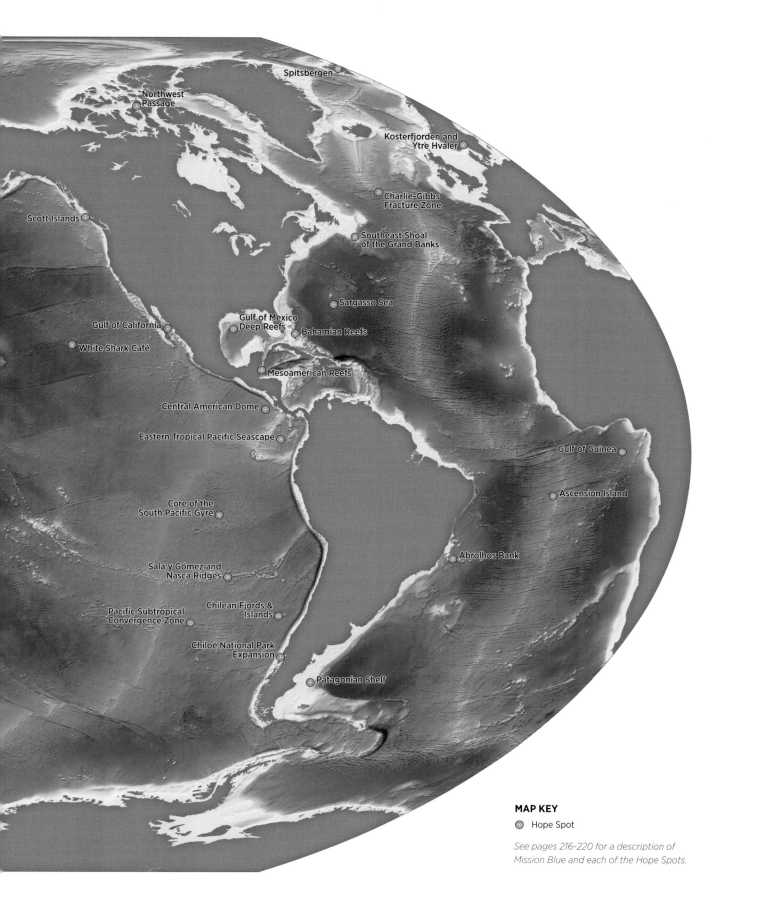

Spitsbergen ◎

Northwest
Passage ◎

Kosterfjorden and
Ytre Hvaler ◎

◎ Charlie-Gibbs
Fracture Zone

Scott Islands ◎

◎ Southeast Shoal
of the Grand Banks

◎ Sargasso Sea

Gulf of California ◎

Gulf of Mexico
Deep Reefs ◎
◎ Bahamian Reefs

White Shark Café ◎

◎ Mesoamerican Reefs

Central American Dome ◎

Gulf of Guinea ◎

Eastern Tropical Pacific Seascape ◎

◎ Ascension Island

Core of the
South Pacific Gyre ◎

◎ Abrolhos Bank

Sala y Gómez and
Nasca Ridges ◎

Chilean Fjords &
Islands ◎

Pacific Subtropical
Convergence Zone ◎

Chiloé National Park
Expansion ◎

◎ Patagonian Shelf

MAP KEY

◎ Hope Spot

*See pages 216-220 for a description of
Mission Blue and each of the Hope Spots.*

WHY
THE
OCEAN
MATTERS

~

With every drop of water you drink,
every breath you take,
you're connected to the ocean,
no matter where on Earth you live.

—SYLVIA A. EARLE

NO BLUE, NO GREEN

THE OCEAN—EARTH'S BLUE HEART

Why should I care about the ocean? I don't swim. I get seasick. People don't drink salt water. If the ocean dried up tomorrow, what difference would it make to me?"

The questions, fired at me by a reporter more than 30 years ago, provoked me to find ways to explain why the ocean matters. "Think about Mars," I said. The beautiful, red planet has lots of rocks, but no blue ocean, no great forests, no birds, fish, or even a single butterfly or bee. The atmosphere there is mostly carbon dioxide.

"If you like to breathe, you will care about the ocean," I continued. After all, more than half of the oxygen in the atmosphere is generated by mostly microscopic marine life that takes up carbon dioxide and water, and through photosynthesis, generates oxygen and sugar. That, in turn, drives great ocean food webs and eventually the chemistry of the biosphere. There may be water without life, but there is no life without water, and most of Earth's water is ocean. No water, no life; no blue, no green.

In a universe where temperatures range from far below water's freezing point to far above the temperature where it boils away as vapor, Earth is blessed with the rare just right temperature range where a liquid ocean is possible. Because of the ocean, life is possible, and over time, Earth's waters have become a living minestrone—every drop a small universe of microbial beings, every cubic mile a thriving metropolis of large, medium, and exquisitely small lives interacting in ways that make Earth hospitable for life as we know it.

Originating in the sea, water forms clouds that return rain, sleet, and snow to the land and ocean, filling lakes and rivers, replenishing groundwater, quenching the thirst of trillions of trees; of ferns, moss, ants, and elephants; of rain forest frogs and desert-dwelling spiders. The water that flows in our blood has cycled time and again from the sea to the sky and back to Earth, where each of us enjoys the use of it for a while. Eventually water returns to the sea, to flow again as an ocean current, as most of the body of a shimmering jellyfish, the moisture in the muscles of a giant tuna, the milk of a mother whale.

It has taken more than three billion years for Earth to be transformed from lifeless rock and water to the complex, vibrant system that now hosts more than seven billion people and many trillions of other animals, plants, fungi, and protists, as well as many more bacteria and other microbes than there are grains of sand. Most of the few million people who existed 10,000 years ago relied on naturally occurring plants and animals for sustenance and materials needed for their survival and well-being, and in so doing, according to Harvard biologist E. O. Wilson, greatly diminished "the large, the slow, the tasty" animals of the land. In North America, thriving populations of woolly mammoths, giant sloths, camels, giant tortoises, and many others animals became extinct thousands of years ago, a pattern replicated on every continent and every island where humans have become established.

SUSAN MIDDLETON ~ MIDWAY ATOLL, NORTH PACIFIC OCEAN
Wisdom, a Laysan albatross banded in the 1950s, warms her most recent egg in the presence of author Sylvia Earle on Midway Atoll. The changes both have witnessed during their shared decades are unprecedented in human—or albatross—history.

Our prosperity as a species has been costly to the natural systems that provide the underpinnings of our existence. While our numbers expanded in a few millennia to a billion by 1800, to twice that by 1930, and twice that again by 1980, ancient forests and grasslands globally were reduced by more than half, and numerous plants, wild birds, reptiles, mammals, amphibians, and invertebrates slipped toward extinction. Thousands of terrestrial species succumbed totally to our increasing dominance.

Owing to our growing numbers and greatly advanced technologies, it has taken only a few centuries, mostly a few recent decades, to have a comparable impact on previously intact ocean ecosystems. Since the 1950s, half of the coral reefs globally have disappeared or are in a state of sharp decline.

Perspective has arrived—just in time. Now we know what no one could understand until recent times: The ocean is Earth's blue engine, driving climate and weather, regulating and stabilizing planetary temperature, governing global chemistry, including the basic cycles of water, oxygen, carbon, nitrogen, phosphorus, and more. Apollo astronaut Edgar Mitchell articulated in 1970 what has now become a vision for all of humankind:

Suddenly, from behind the rim of the Moon, in long, slow-motion moments of immense majesty, there emerges a sparkling blue and white jewel, a light delicate sky-blue sphere laced with slowly swirling veils of white, rising gradually like a small pearl in a thick sea of black mystery. It takes more than a moment to fully realize this is Earth . . . home.

The ocean drives the water cycle, climate, and weather; stabilizes temperature, generates most of the oxygen in the atmosphere, takes up much of the carbon dioxide, shapes planetary chemistry, holds the planet steady.

—SYLVIA A. EARLE

In the Mediterranean and Caribbean Seas, 80 percent are gone. Ninety percent of many kinds of fish have been eliminated, following the "large, slow, and tasty" depletion of land animals, but also including the small (herring, capelin, shrimp, oysters, clams, and many others), the swift (tuna, swordfish, marlin), and the notably unpalatable (menhaden, krill, horseshoe crabs).

The idea of terraforming Mars to make our sister planet more Earthlike is gaining traction at the same time that our collective actions are *marsiforming* Earth, making the blue planet more like the red one. But like a skier who suddenly becomes aware of a steep cliff ahead, we can now see what our predecessors could not: There are limits that must be respected if we are to survive.

About half the face of that jewel is occupied by the Pacific Ocean, Earth's oldest, deepest, widest, and wildest body of water. In 2012, I made a pilgrimage to one of the tiny specks of land embraced by that immense expanse of ocean, Midway Atoll, part of the Northwestern Hawaiian Islands located about halfway between San Francisco and Tokyo. Six years earlier I had stood by the side of President George W. Bush as he signed into law the designation of the atoll and dozens of other islands and islets within a 140,000-square-mile area of land and sea as the Papahānaumokuākea Marine National Monument, now a permanent safe haven for life on the land and in the surrounding sea. In a stroke of the president's pen,

protection was granted to places of cultural significance to native Hawaiians, and to the continued existence of birds, seals, fish, and other wildlife of the area. For me, it is a treasured Hope Spot, one of more than 50 special places in the sea now recognized as areas where safeguards will have a magnified impact on the health of the ocean, and therefore provide hope for maintaining vital pristine seas and restoring damaged systems.

It took special permission for me to be allowed to dive along the outer wall of the atoll, swim among dozens of sharks and hundreds of reef fish and later, to walk among tens of thousands of nesting seabirds. Serenely indifferent to the approach of three individuals many times her size, Wisdom, a well-known Laysan albatross, acknowledged our presence with a slight tilt of her head. I nodded in return, a small gesture of respect for the grand dame of Midway Atoll, a 62-year-old mother bird warming her single egg of the year within an orderly cushion of grass blades and downy feathers. Wisdom's lifetime mate was at sea, but would likely be back in time to greet their miraculously morphed hatchling, only weeks before a mass of gooey yolk and slippery albumen, soon to emerge from its ivory-shelled incubator as a gangly but complete bird—heart, muscle, lungs, bone, brain, incipient feathers, and exceedingly bright eyes.

A hatchling herself in 1950, Wisdom was learning to fly at about the same time I took to the sea, learning to dive. For her, mastering the wind was essential to being who she was, the distillation of thousands of generations of long-winged creatures, fine-tuned to stay aloft for months with aeronautic efficiency that makes even the most learned of engineers sigh with envy. For me, mastering access to the sea was more choice than necessity, but in fact, I found the ocean to be totally irresistible long before I left the New Jersey nest of my youth. Flying over thousands of miles of ocean every year, Wisdom has witnessed in her lifetime changes that are unprecedented

DAVID LIITTSCHWAGER ~ NORTHWESTERN HAWAIIAN ISLANDS, NORTH PACIFIC OCEAN
Many hatchling seabirds never live to fly owing to ingested plastics, the apparent fate of this Laysan albatross chick at Kure Atoll in the Northwestern Hawaiian Islands.

in the millions of years that albatrosses have existed. Diving thousands of hours under the sea, I have witnessed similar changes in the depths below.

But as skilled as she is, Wisdom cannot know why it has become increasingly more difficult to find squid and small fish to eat, or understand the source of the avalanche of plastics and other trash in the sea, or know the danger of baited hooks on lines 50 miles long that were not there when she began to fly. Even if she did know why, she is powerless to do anything about it.

We do know what is driving the changes that are altering the nature of the ocean, and we do have the power to reverse the unfavorable trends. Some actions are under way. Technologies that have taken us high in the sky, deep in the sea, and connected the thoughts of billions of minds are yielding insights that no others before us have been able to grasp. Now we know. We must take care of the ocean as if our lives depend on it, because they do.

"Hope" is the thing
with feathers
That perches in the soul,
And sings the tune
without the words,
And never stops at all

—EMILY DICKINSON, POET

**FRANS LANTING ~ HAWAIIAN LEEWARD ISLANDS,
NORTH PACIFIC OCEAN**
*Time for takeoff! Juvenile Laysan albatrosses
prepare for a lifetime of long-distance flight from
their home nests on the Hawaiian Leeward Islands.*

FRANS LANTING ~ SOUTH PACIFIC OCEAN
(PRECEDING PAGES)
*Water as vapor clouds the sky over the vast
Pacific Ocean—Earth's largest, deepest, widest,
and oldest body of water.*

We are dependent on the
health of the ocean at the
most fundamental level.
If you protect the ocean,
you protect yourself.

—JEAN-MICHEL COUSTEAU,
AMERICA'S UNDERWATER TREASURES

FLIP NICKLIN ~ HAWAIIAN ISLANDS,
NORTH PACIFIC ISLAND
*Minds in the sea: A pod of bottlenose
dolphins, marine mammals known for
their intelligence and highly social nature,
sweeps through Hawaiian waters.*

CHRIS NEWBERT ~ SOLOMON ISLANDS,
SOUTH PACIFIC OCEAN

Like its jellyfish cousins, this corallimorpharian
in the Solomon Islands has soft tentacles
surrounding the mouth.

DAVID DOUBILET ~ INDONESIA,
SOUTH PACIFIC OCEAN

Delicate but durable, diaphanous moon jellies
pulse in quiet seas near Raja Ampat in the
South Pacific.

Through the cycling of water, across space and time,
we are linked to all life . . . Water's gift is life.
No water, no life.

—SANDRA POSTEL, FOUNDER, GLOBAL WATER POLICY PROJECT

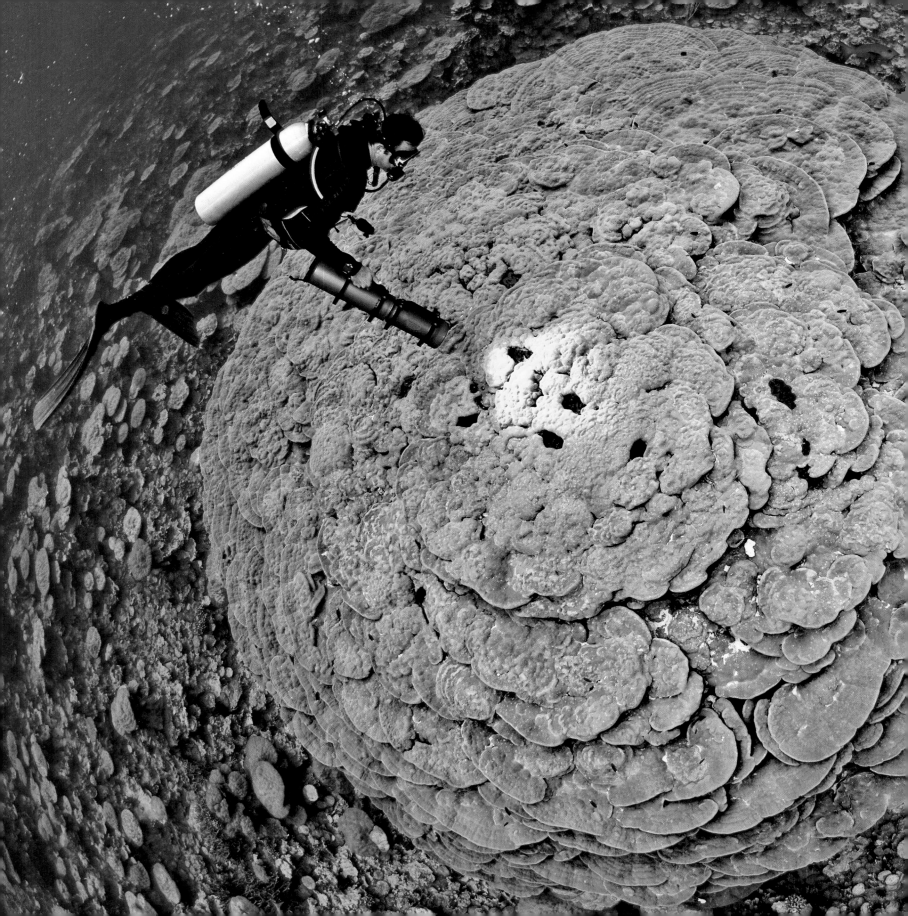

We know that when
we protect our oceans
we're protecting
our future.

—PRESIDENT BILL CLINTON

BRIAN SKERRY ~ KINGMAN REEF, NORTH PACIFIC OCEAN
*Cause for hope: Thousands of individual animals
connect to form massive coral colonies at Kingman Reef,
a pristine area in the Northern Line Islands, a national
wildlife refuge area.*

CHRIS NEWBERT ~ SOLOMON ISLANDS, SOUTH PACIFIC OCEAN
Daisy-like coral polyps share an abandoned barnacle shell with a hermit crab wielding distinctive peppermint-patterned pinchers.

CHRIS NEWBERT ~ MICRONESIA, WESTERN PACIFIC OCEAN
Millions of years before dinosaurs existed, feather stars and gorgonians snared passing plankton in tropical seas—and they are still doing it near the island nation of Palau.

The ocean is a wilderness reaching round the globe,
wilder than a Bengal jungle, . . . washing the very
wharves of our cities and the gardens
of our seaside residences.

—HENRY DAVID THOREAU, *CAPE COD*

To me the sea
is a continual miracle,
The fishes that swim—
the rocks—the motion
of the waves—the ships
with men in them,
What stranger
miracles are there?

—WALT WHITMAN, "MIRACLES," *LEAVES OF GRASS*

MATTIAS KLUM ~ FIJI, SOUTH PACIFIC OCEAN
Clinging to its fast-moving host, a tiny remora finds safe transport aboard a whitetip reef shark. Assorted reef fish frequent the waters of Shark Reef near Beqa Island.

MICHAEL AW ~ CELEBES SEA, WESTERN PACIFIC OCEAN
A whimsically ornate acrocirrid worm, cousin of terrestrial earthworms, lives in darkness deep within the Celebes Sea.

DAVID WROBEL ~ EASTERN PACIFIC OCEAN
More than 2,000 feet deep in the eastern Pacific Ocean, elegantly graceful creatures such as this belie the name bestowed by scientists: dogface witch eels.

Wildness reminds us what it means to be human,
what we are connected to rather than
what we are separate from.

—TERRY TEMPEST WILLIAMS, AUTHOR & CONSERVATIONIST

Northwestern Hawaiian Islands

Kure Atoll (Ocean I.)
Midway Islands
Pearl and Hermes Atoll
Lisianski I.
Laysan I.
Gardner Pinnacles
La Perouse Pinnacle
Necker I.
Nihoa
Ni'ihau
Kaua'i
O'ahu
Moloka'i
Maui
Lāna'i
Kaho'olawe
Hawai'i

Papahānaumokuākea Marine National Monument

0 km   200
0 mi   200

PAPAHĀNAUMOKUĀKEA MARINE NATIONAL MONUMENT

Designated by President George W. Bush as a national underwater monument in 2006, these northwestern Hawaiian Islands encompass 139,797 square miles (362,073 km), one of the largest fully marine protected areas on Earth and larger than all of the U.S. national parks combined. The reefs and surrounding sea provide home for more than 7,000 marine species and 22 breeding and nesting seabirds. About a quarter of the plants and animals there are endemic to the Hawaiian Islands and include several endangered species such as monk seals and Laysan ducks.

PACIFIC OCEAN

The blue face of the Earth when viewed from space, the Pacific is the oldest, widest and deepest ocean encompassing 58,925,815 square miles (152,617,159 sq km), nearly half of the planet's water, averaging about 13,800 feet (4,200 m) in depth, with a maximum of 35,787 feet (10,908 m) in the Challenger Deep, confirmed from a submersible on-site in 2012 by James Cameron. Edges of the underlying Pacific Plate form the notorious "Ring of Fire," home to 70 percent of the world's earthquakes, most of its volcanic eruptions, and most of the world's inhabited islands. The highest mountain in the world, Mauna Kea, rises from the seafloor to a towering 33,100 feet (10,100 m), with about 13,800 feet (4,200 m) extending above the ocean's surface.

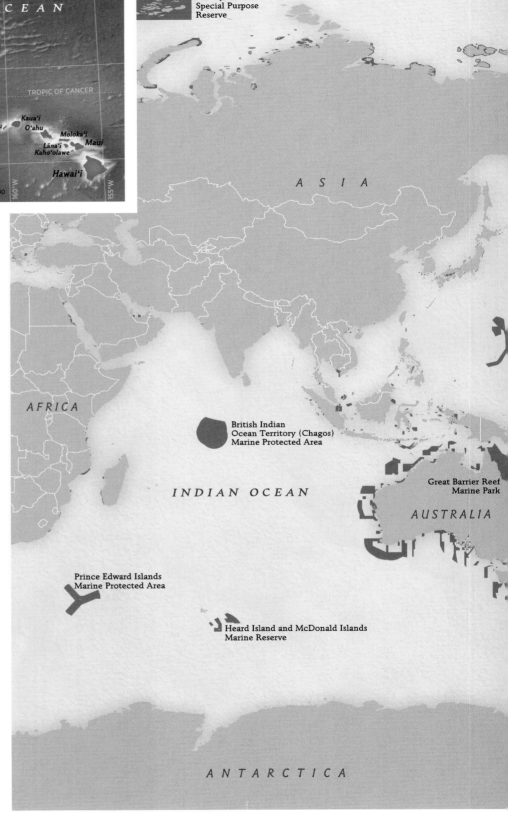

Franz Josef Land
Special Purpose
Reserve

ASIA

AFRICA

British Indian
Ocean Territory (Chagos)
Marine Protected Area

Great Barrier Reef
Marine Park

INDIAN OCEAN

AUSTRALIA

Prince Edward Islands
Marine Protected Area

Heard Island and McDonald Islands
Marine Reserve

ANTARCTICA

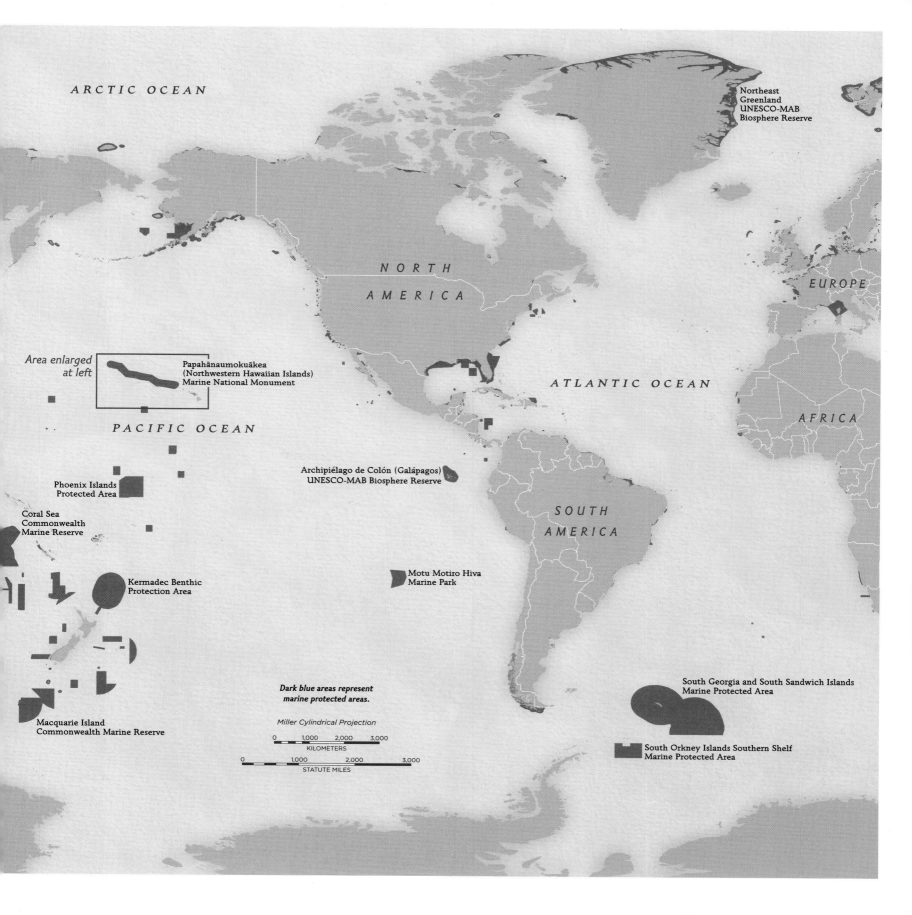

ARCTIC OCEAN

Northeast
Greenland
UNESCO-MAB
Biosphere Reserve

NORTH
AMERICA

EUROPE

Area enlarged
at left

Papahānaumokuākea
(Northwestern Hawaiian Islands)
Marine National Monument

ATLANTIC OCEAN

AFRICA

PACIFIC OCEAN

Phoenix Islands
Protected Area

Archipiélago de Colón (Galápagos)
UNESCO-MAB Biosphere Reserve

SOUTH
AMERICA

Coral Sea
Commonwealth
Marine Reserve

Kermadec Benthic
Protection Area

Motu Motiro Hiva
Marine Park

South Georgia and South Sandwich Islands
Marine Protected Area

Dark blue areas represent
marine protected areas.

Macquarie Island
Commonwealth Marine Reserve

Miller Cylindrical Projection

0   1,000   2,000   3,000
KILOMETERS

0   1,000   2,000   3,000
STATUTE MILES

South Orkney Islands Southern Shelf
Marine Protected Area

DIVING

INTO

THE

BLUE

~

It was easy to forget I was supposed
to be a serious scientist
as I careened around . . .
absorbing and savoring every
new image of discovery
made sensuous by the flow of a
warm silken sea against my bare skin.

—SYLVIA A. EARLE

BECOMING ONE WITH THE OCEAN

THE GULF OF MEXICO & THE CARIBBEAN SEA

As a fearless three-year-old, I was tumbled by a rogue wave of green water. Cresting white, the Atlantic Ocean engulfed me, sweeping me upside down, pulling me deeper, holding me, then shifting . . . surged me back to where my toes touched sand and strong arms lifted me to safety. A moment later, I jumped back in, exhilarated.

Fifteen years later, committed to doing whatever it took to become a scientist, I fastened the straps of an air-filled cylinder to my back, bit down on a black rubber mouthpiece, slipped into the Gulf of Mexico, and focused on my two words of instruction: "Breathe naturally."

What must it be like to be a bird, for the first time lifting off, rising above Earth previously so solidly there, now viewed merely as a platform to launch from or return to for a while. Underwater, I imagined my arms were wings, and improbable as it seems, I could fly! And breathe! Perhaps young birds feel similar disbelief during their first days of gliding above the ground, experiencing sensations of wonder that may not be uniquely human.

The ocean pioneer Jacques Cousteau related the joy of being weightless in his first book, *The Silent World:* "At night I had often had visions of flying . . . now I flew without wings." Mesmerized by Cousteau's writings and films of the 1950s, I, too, wanted to explore the clear blue Mediterranean Sea, glittering with hundreds of kinds of fish, forested with angular corals, giant sponges, and rainbow-colored plankton. From my home in Dunedin, Florida, where my parents moved the family when I was 12, my "Mediterranean" was the Gulf of Mexico, similarly populated with several kinds of sea turtles, grouper, snapper, coral reefs, sea grass meadows, and shallow mangrove-fringed coastal areas in places sloping off into dark waters thousands of feet deep.

I was entranced with the Gulf's counterpart to the Mediterranean dugong, Florida manatees that graze like pudgy cows on sea grass meadows and sometimes bask in Florida's freshwater rivers and springs. At the time, I was not aware that another kind of marine mammal, monk seals, once ranged widely throughout the Gulf and Caribbean, lounging on beaches as far north as Galveston, Texas. Jacques Cousteau encountered their cousins in the Mediterranean Sea, and I have met their Hawaiian relatives at Midway Island, but the last playful, whiskered monk seal face was seen in the Gulf in 1952.

The number of manatees in the Gulf and Caribbean Sea presently are a fraction of their former abundance, and the same is true of many of the more than 15,000 other forms of life—whales, sea turtles, shrimp, spiny lobsters, stone crabs, blue crabs, oysters, clams, conchs, sharks, rays, and other fish as well as mangroves, sea grasses, salt marshes, and even microscopic plankton and numerous invertebrates that over eons have shaped the region into one of the most productive bodies of water on Earth. While life in the ocean has markedly decreased, one species—humankind—has dramatically

KIP EVANS ~ FLORIDA KEYS, NORTH ATLANTIC OCEAN
In 2012, Sylvia Earle peeks from the porthole of Aquarius in 60 feet of water. Earle submerged for a week with five fellow aquanauts to explore Conch Reef in the Florida Keys National Marine Sanctuary, using the undersea lab as a home base. This marked her tenth experience living under the sea.

increased, our numbers more than doubling globally and by tenfold in Florida since I first dived into the Gulf of Mexico.

Places I knew and loved as a child began to change rapidly in the 1960s, as marshes and mangroves were transformed into marinas, industrial sites, and housing developments. I suggested that Clearwater High School where I graduated in 1952 should be required to change its name to something like "Murkywater High" to reflect reality. But offshore and much of Florida's west coast remained wild during years when I explored the Gulf Coast from the southernmost point of land in the country, the Dry Tortugas Islands, to the Mississippi Delta, diving and documenting the hundreds of kinds of marine algae and the thousands of animals that lived on and among their

scientist, though, and could not resist responding to a request for applicants who wanted to conduct research while living submerged for two weeks in an underwater laboratory in the U.S. Virgin Islands.

Sponsored by the Navy, NASA, the Department of the Interior, the Smithsonian Institution, and the General Electric company, Project Tektite in 1970 would provide ten five-person teams—four scientists and one engineer—a chance to use the ocean as a laboratory when not inside the warm, dry living quarters 50 feet under the sea, eating, sleeping, taking hot showers, and writing reports. No women had been accepted as astronauts at the time, but there was no mention of gender on the application to become an *aquanaut*.

When people ask where is the best place to go diving,
I say either, "The next place—wherever that is!"
or "Almost anywhere, 50 years ago."
—SYLVIA A. EARLE

lacy fronds and translucent pink, green, and golden brown branches. Often, I enlisted my two young children, Elizabeth and Richie, to help prepare museum specimens that today provide tangible evidence of the existence of undersea forests and meadows that have now disappeared.

Years before formal classes in scuba diving were available, I sometimes tagged along with tolerant U.S. Navy divers in Panama City, Florida, for research dives to two platforms located 2 and 12 miles offshore and, by watching the experts, I learned a lot about the dos and don'ts of staying alive underwater.

By 1969, as a research fellow at Harvard University, I had logged more than a thousand hours underwater, but had not yet been officially certified as a diver. I was a "certified"

Despite the consternation of some, the head of the program, James Miller, decided to assemble from qualified applicants a team of women (mixed teams were considered unthinkable), astutely noting, "Half the fish are female. Half the dolphins and whales. I think women will do ok." More than "ok," women and men alike experienced unprecedented time, day and night, exploring the reefs and gaining new insight concerning the nature of the complex metropolis of sea life in Lameshur Bay, St. John.

As a scientist, I already understood that no two fish or snails or squids are exactly alike, in the same way that each dog, cat, horse, or human is unique, but I had never before become so well acquainted with individual fish that I could recognize faces and behavior that distinguished members of the same species. At dawn, I watched five young gray angelfish emerge from their separate overnight sleeping quarters and gather as a group to prowl the reef, pausing from time to time to graze on choice bits of seaweed or nibble a tasty sponge. Barracuda—long, sleek animals with big teeth and an attitude to match—at first glance seemed to all be alike, but looking closely, the faces were different, with varied dark spots on their silvery flanks. Some were more curious than others, some more reticent. Having often seen fish swimming in butter on a dinner plate, it was a revelation to me to view them with new eyes, and respect not only their individuality but also their collective roles in the reef ecosystem. I began to understand that a coral reef functions not just through the existence of corals, any more than buildings make a city or trees a forest.

As a serious scientist, I had spent years learning everything I could about the nature of the ocean, about the creatures who live there, and about new technologies that make it possible to live underwater and move silently through the ocean with advanced rebreather systems comparable to what skywalking astronauts wear in space. But nothing prepared me for the public reaction to the news that a team of women was living under the sea. When we emerged, hundreds of reporters with microphones and cameras splashed news around the world about the adventures of the "aquababes," "aquabelles," even the "aquanaughties," and we were plied with urgent questions such as "Did you wear lipstick?" "Did you have a hair drier?" We were given keys to the city and a ticker-tape parade down State Street in Chicago, feted at a White House luncheon, and given medals by the Department of the Interior. As team leader, I was chosen to report to Congress about the results of our mission.

Bewildered at first, I soon realized that people really were interested in what it was like to live underwater, and I might be able to find ways to build their interest and

JOEL SARTORE ~ GULF OF MEXICO
Plumes of smoke unfurl over massive amounts of oil set afire in April 2010. The oil gushed from a mile underwater into the Gulf of Mexico from the broken Macando Deepwater Horizon wellhead.

curiosity into a better understanding of the ocean—and why caring for the ocean matters. Yet, when the National Geographic Society asked me to write about the experience for its prestigious magazine, my first reaction was no! Scientists write for other scientists, not for the public, I thought. But then, I thought again. I had just explored parts of the universe within our home planet that even high-flying astronauts had not witnessed. How could I not share what I had seen, felt, and come to understand?

The August 1971 issue of *National Geographic* magazine was a turning point. Since then, I have lived underwater nine more times, most recently during the summer of 2012, in the Aquarius Underwater Laboratory in the Florida Keys, fondly known as "America's Space Station in the Sea." Again, there were headlines, but this time the focus was on how much the reef has recovered owing to protection during the 20 years that Aquarius has been in its present location. Almost no one remarked on the fact that I spent a week submerged with five handsome aquahunks.

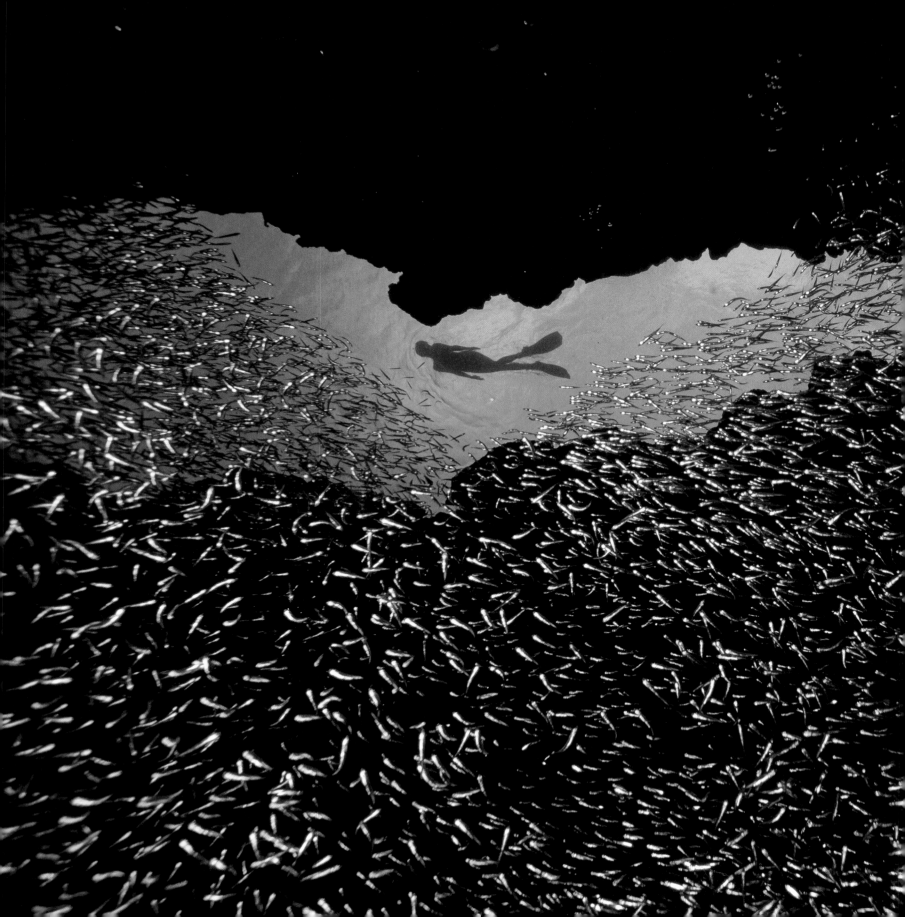

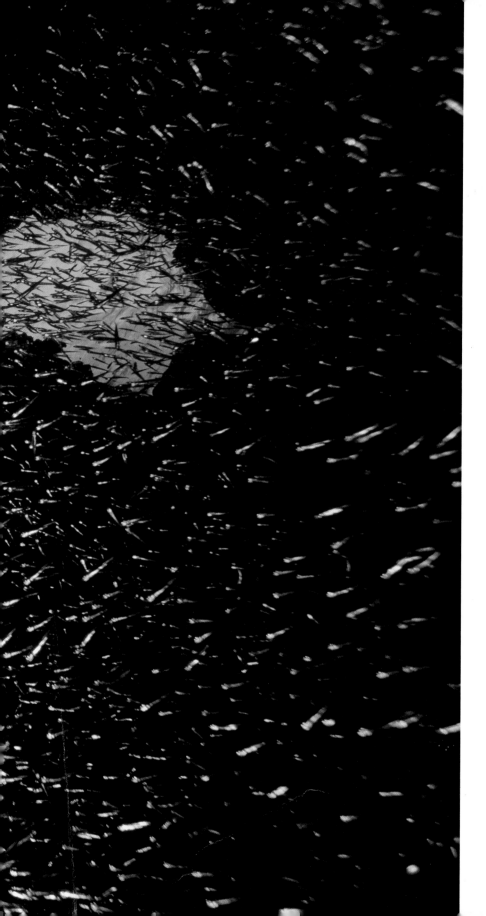

Let me take you where I go
Suspended in infinite grace
For one moment of
perfect beauty
In the ocean's warm embrace
Let me take you by the hand
In an endless liquid sky
Let me take you where I go
And together we will fly.

—GALE MEAD, MUSICIAN

DAVID DOUBILET ~ WEST INDIES, CARIBBEAN SEA
A glittering veil of fish surrounds a diver in the Cayman Islands—peaks of a massive underwater ridge known as the Cayman Ridge.

BRIAN SKERRY ~ BELIZE, CARIBBEAN SEA
(PRECEDING PAGES)
Like a great blue eye, the Blue Hole gleams among reefs near Lighthouse Reef. Ninety feet below, stalactites and stalagmites circle the walls of what is thought to be an ancient sinkhole formed when sea level was lower.

The world, we are told, was
made especially for man—
a presumption not supported
by all the facts.

—JOHN MUIR, *A THOUSAND MILE WALK TO THE GULF*

BRIAN SKERRY ~ BELIZE, CARIBBEAN SEA
*Hol Chan Marine Reserve, the first marine
park in Belize, is home for a school of
schoolmaster snappers.*

But more wonderful
than the lore of old men
and the lore of books
is the secret lore of ocean.

—H. P. LOVECRAFT, "THE WHITE SHIP"

BRIAN SKERRY ~ BELIZE, CARIBBEAN SEA
Endowed with oversize curiosity and hearty
appetites, black groupers such as this are
treasured by divers and are vital to the health
of Caribbean reef systems.

PAUL SUTHERLAND ~ NETHERLANDS ANTILLES, CARIBBEAN SEA
Crowned with feathery feeding filters, a Christmas tree worm dines on plankton near Bonaire Island in the Caribbean Sea.

DAVID DOUBILET ~ EAST FLOWER GARDEN BANK, GULF OF MEXICO
A steel archipelago of oil and gas platforms in the northern Gulf of Mexico shelters thousands of creatures such as this tessellated blenny living in Flower Garden Banks National Marine Sanctuary.

What magic the underwater world holds!
There are creatures that are speckled, spotted,
see through, and those that glow in the dark.

—DARYL HANNAH, ACTRESS & ENVIRONMENTALIST

Thousands have lived
without love;
not one without water.

—W. H. AUDEN, "FIRST THINGS FIRST"

BRYCE GROARK ~ GULF OF MEXICO
Whale sharks, the biggest fish in the sea, have patterns of spots as distinctive as fingerprints. The patterns are used to track movements of individuals such as this graceful giant feeding on fish eggs in the Gulf of Mexico, with Sylvia Earle looking on.

We need another
and a wiser
and perhaps
a more mystical concept
of animals.
. . . They are not brethren,
they are not underlings;
they are other nations,
caught with ourselves
in the net of life and time.

—HENRY BESTON, *THE OUTERMOST HOUSE*

SANDRA CRITELLI ~ GULF OF MEXICO
Massive migrations of cownose rays
periodically fly like giant, golden moths
across the Gulf of Mexico.

It is up to us to conserve
the most important
wild areas that remain.
Doing so will preserve
something that is
all too easy to destroy
but impossible to replace.

—LAURA BUSH, FORMER FIRST LADY

PAUL NICKLEN ~ FLORIDA, GULF OF MEXICO
*Manatees, once abundant throughout the Gulf
and Caribbean Sea, have hope for recovery in
Kings Bay, the headwater of Crystal River, a
national wildlife refuge on Florida's west coast.*

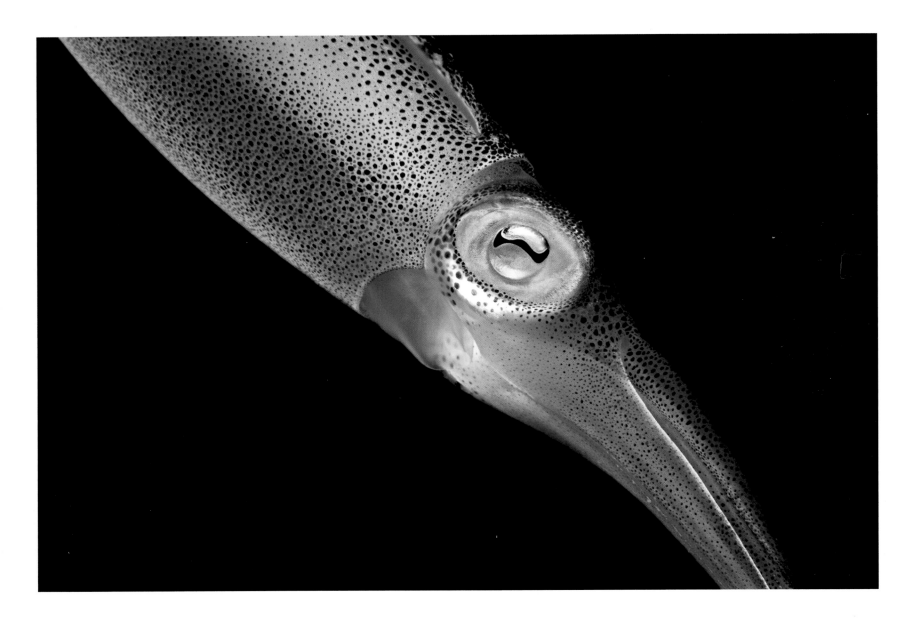

DAVID FLEETHAM ~ CARIBBEAN SEA
*Related to oysters, clams, and octopods,
this Caribbean reef squid has eyes remarkably
similar in structure to those of vertebrates.*

BRIAN SKERRY ~ CARIBBEAN SEA
*A delicate membrane encloses this developing
embryonic Caribbean reef squid in a protected
liquid space.*

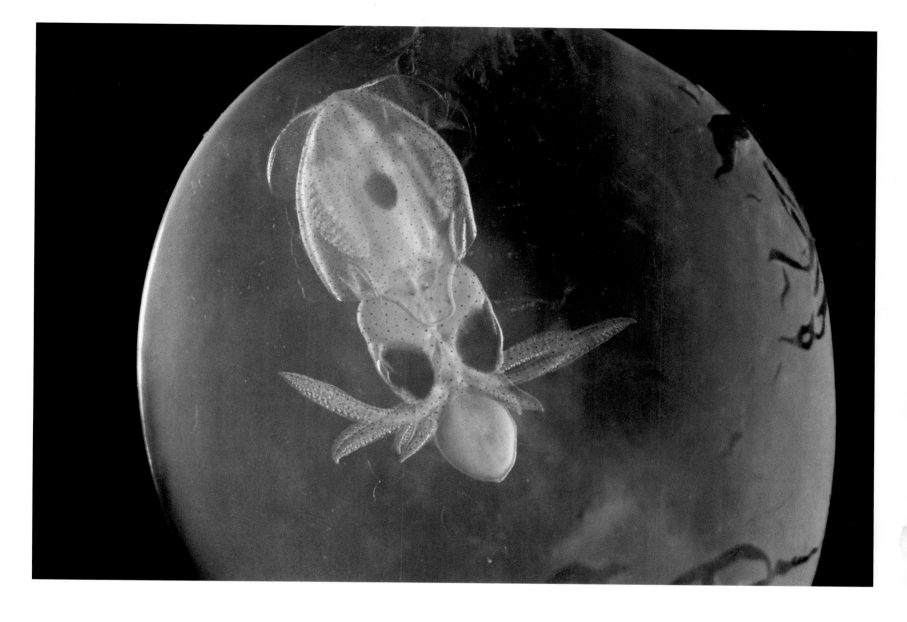

If there is magic on this planet,
it is contained in water.

—LOREN EISELEY, *THE IMMENSE JOURNEY*

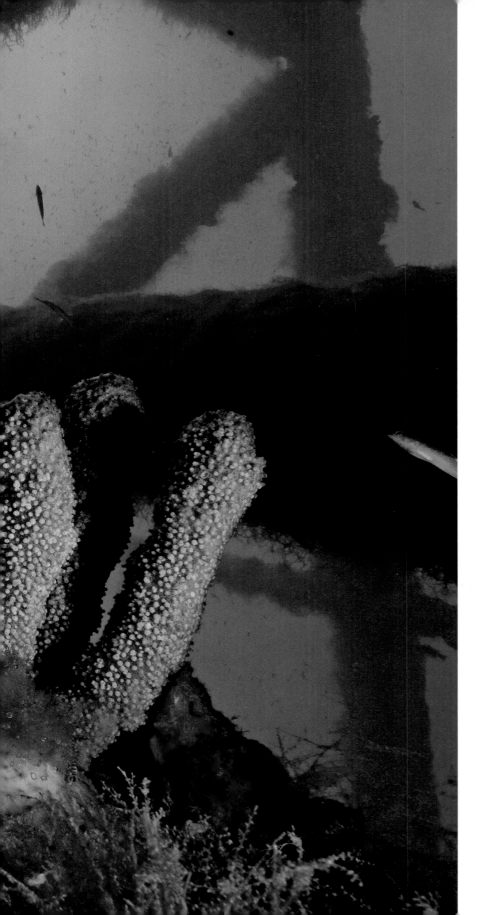

In the end, our society
will be defined not only by
what we create, but by what
we refuse to destroy.

—JOHN SAWHILL,
PRESIDENT OF THE NATURE CONSERVANCY, 1990–2000

DAVID DOUBILET ~ GULF OF MEXICO
*A cluster of tube sponges clings to the sturdy
pillars of an oil rig in the northern Gulf of
Mexico, an attraction for recreational divers.*

DAVID SHALE ~ GULF OF MEXICO
*Iridescent bands of cilia propel this lobate comb
jelly through deep water in the Gulf of Mexico.*

DAVID SHALE ~ GULF OF MEXICO
*With the size, shape, and consistency of a translucent plum, this gelatinous
squid feeds while suspended in darkness in deep water in the Gulf of Mexico.*

This strange world . . . will remain forever
the most vivid memory . . . its cosmic chill
and isolation, the eternal and absolute darkness
and the indescribable beauty of its inhabitants.

—WILLIAM BEEBE, *HALF MILE DOWN*

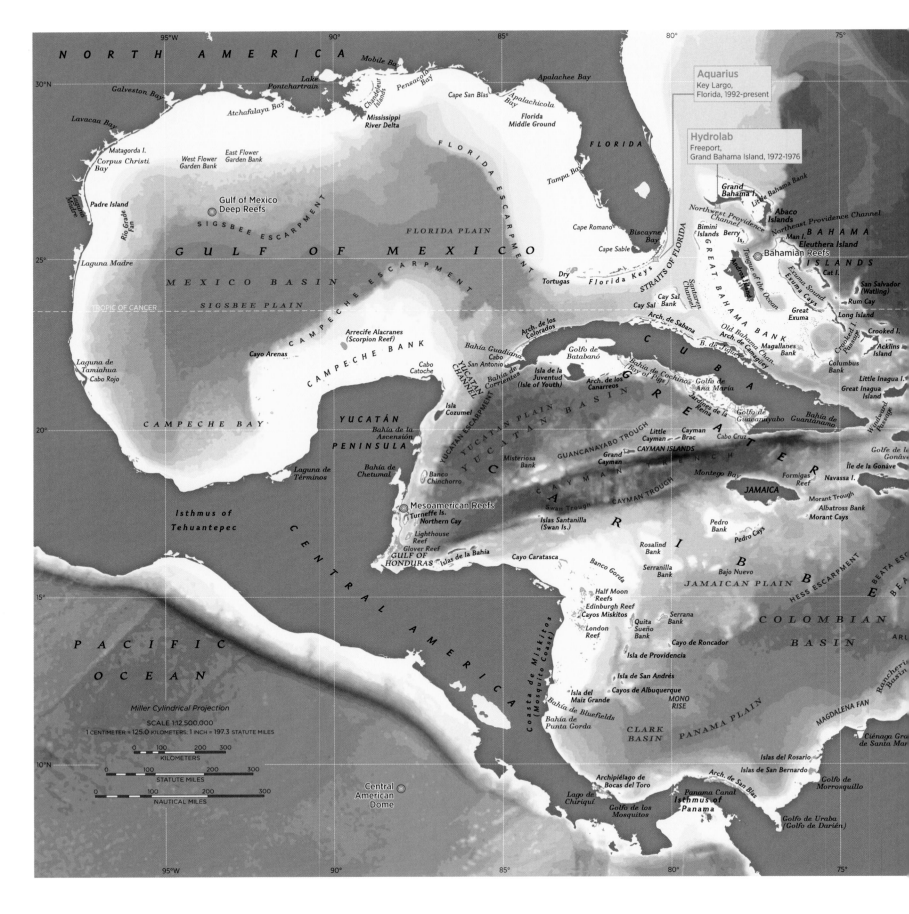

NORTH AMERICA

Mobile Bay
Lake Pontchartrain
Galveston Bay
Atchafalaya Bay
Pensacola Bay
Chandeleur Islands
Cape San Blas
Apalachee Bay
Apalachicola Bay
30°N
Lavaca Bay
Mississippi River Delta
Florida Middle Ground
FLORIDA
Matagorda I.
Corpus Christi Bay
West Flower Garden Bank
East Flower Garden Bank
Tampa Bay

Aquarius
Key Largo, Florida, 1992–present

Hydrolab
Freeport, Grand Bahama Island, 1972–1976

Padre Island
Gulf of Mexico Deep Reefs
Grand Bahama I.
Little Bahama Bank
Laguna Madre
Abaco Islands
Northwest Providence Channel
Northeast Providence Channel
Rio Grande Fan
SIGSBEE ESCARPMENT
Bimini Islands
Berry Is.
Man I.
BAHAMA
Eleuthera Island

25°
Laguna Madre
GULF OF MEXICO
FLORIDA PLAIN
Cape Romano
Biscayne Bay
Andros Island
Bahamian Reefs
ISLANDS
Cat I.

MEXICO BASIN
Cape Sable
Dry Tortugas
Florida Keys
Straits of Florida
Cay Sal Channel
Santaren Channel
Tongue of the Ocean
Exuma Sound
Exuma Cays
San Salvador (Watling)

TROPIC OF CANCER
SIGSBEE PLAIN
Cay Sal Bank
Cay Sal
Arch. de Sabana
Great Exuma
Rum Cay
Long Island

Arrecife Alacranes (Scorpion Reef)
Arch. de los Colorados
Golfo de Batabanó
B. de Jigüey
Old Bahama Chan.
Arch. de Camagüey
Magallanes Bank
Crooked I.
Columbus Bank
Crooked Island
Acklins Island

Cayo Arenas
CAMPECHE BANK
Cabo Catoche
Bahía Guadiana
Cabo San Antonio
Bahía de Corrientes
Isla de la Juventud (Isle of Youth)
Bahía de Cochinos (Bay of Pigs)
Golfo de Ana María
C U B A
Jardines de la Reina
Golfo de Guacanayabo
Bahía de Guantánamo
Little Inagua I.
Great Inagua Island

20°
Isla Cozumel
YUCATÁN PENINSULA
Bahía de la Ascensión
Misteriosa Bank
Little Cayman
Cayman Brac
Grand Cayman
CAYMAN ISLANDS
Golfo de Guacanayabo
Cabo Cruz
Windward Passage
Golfe de la Gonâve
Île de la Gonâve
Navassa I.

CAMPECHE BAY
Bahía de Chetumal
Banco Chinchorro
Laguna de Términos
GUANCANAYABO TROUGH
Cayman TRENCH
Montego Bay
Formigas Reef
Morant Trough
Albatross Bank
Morant Cays

Isthmus of Tehuantepec
Mesoamerican Reefs
Turneffe Is.
Northern Cay
Lighthouse Reef
Glover Reef
CAYMAN TROUGH
Swan Trough
JAMAICA
JAMAICAN PLAIN
HESS ESCARPMENT

GULF OF HONDURAS
Islas de la Bahía
Cayo Caratasca
Islas Santanilla (Swan Is.)
Rosalind Bank
Banco Gorda
Pedro Bank
Pedro Cays
Bajo Nuevo
COLOMBIAN BASIN
BEATA ESC

15°
Half Moon Reefs
Edinburgh Reef
Cayos Miskitos
London Reef
Serranilla Bank
Quita Sueño Bank
Serrana Bank
Cayo de Roncador

PACIFIC OCEAN
Isla de Providencia
Isla de San Andrés
Cayos de Albuquerque
MONO RISE
ARU

CENTRAL AMERICA
Costa de Miskitos (Mosquito Coast)
Isla del Maíz Grande
Bahía de Bluefields
Bahía de Punta Gorda
CLARK BASIN
PANAMA PLAIN
MAGDALENA FAN
Rancheria Basin

Miller Cylindrical Projection
SCALE 1:12,500,000
1 CENTIMETER = 125.0 KILOMETERS; 1 INCH = 197.3 STATUTE MILES

0 100 200 300
KILOMETERS

0 100 200 300
STATUTE MILES

0 100 200 300
NAUTICAL MILES

Central American Dome
Archipiélago de Bocas del Toro
Lago de Chiriquí
Golfo de los Mosquitos
Panama Canal
Isthmus of Panama
Arch. de San Blas
Islas del Rosario
Islas de San Bernardo
Golfo de Morrosquillo
Golfo de Uraba (Golfo de Darién)
Ciénaga Gra de Santa Mar

10°N

95°W
90°
85°
80°
75°

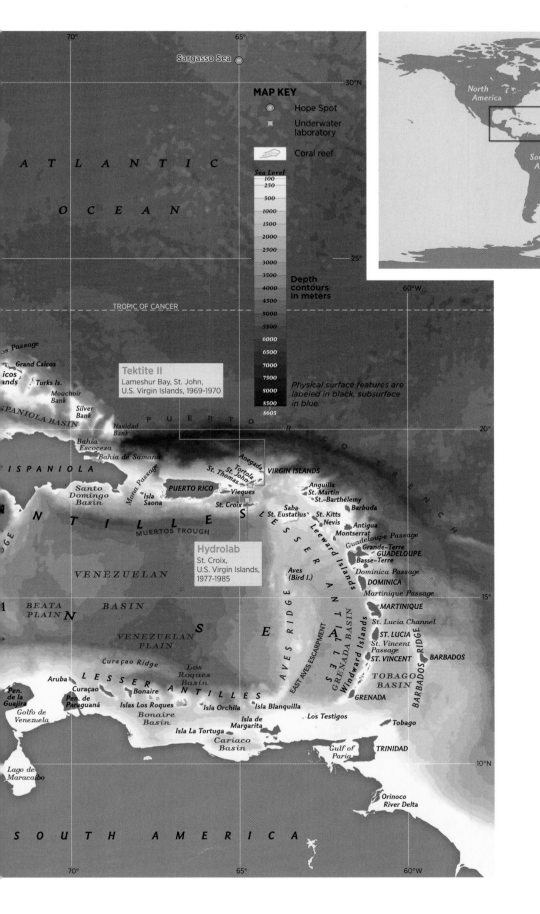

Sargasso Sea ◎

MAP KEY
◎ Hope Spot
✦ Underwater laboratory
〰 Coral reef

Sea Level
100
250
500
1000
1500
2000
2500
3000
3500
4000
4500
5000
5500
6000
6500
7000
7500
8000
8500
8605

Depth contours in meters

Physical surface features are labeled in black, subsurface in blue.

A T L A N T I C

O C E A N

TROPIC OF CANCER

Tektite II
Lameshur Bay, St. John, U.S. Virgin Islands, 1969-1970

os Passage
Grand Caicos
icos Turks Is.
ands Mouchoir Bank
Silver Bank
SPANIOLA BASIN
Navidad Bank

Bahía Escocesa
Bahía de Samaná
P U E R T O   R I C O   T R E N C H

Anegada
Tortola
St. John
St. Thomas
VIRGIN ISLANDS
Anguilla
St. Martin
St.-Barthélemy
Barbuda

ISPANIOLA
Santo Domingo Basin
Mona Passage
Isla Saona
PUERTO RICO
Vieques
St. Croix
Saba
St. Eustatius
St. Kitts
Nevis
Montserrat
Antigua
Guadeloupe Passage
Grande-Terre
GUADELOUPE
Basse-Terre

NTILLES
MUERTOS TROUGH

Hydrolab
St. Croix, U.S. Virgin Islands, 1977-1985

VENEZUELAN

Aves (Bird I.)

Dominica Passage
DOMINICA
Martinique Passage
MARTINIQUE
St. Lucia Channel
ST. LUCIA
St. Vincent Passage
ST. VINCENT
BARBADOS

BEATA PLAIN
N
BASIN
S

VENEZUELAN PLAIN

Curaçao Ridge
Los Roques Basin
Aruba
LESSER ANTILLES
Curaçao
Bonaire
Islas Los Roques
Isla Orchila
Isla Blanquilla
Los Testigos
Tobago

Pen. de la Guajira
Pen. de Paraguaná
Bonaire Basin
Isla La Tortuga
Isla de Margarita
Cariaco Basin
Gulf of Paria
TRINIDAD

Golfo de Venezuela

Lago de Maracaibo

Orinoco River Delta

S O U T H   A M E R I C A

AVES RIDGE
EAST AVES ESCARPMENT
Leeward Islands
Windward Islands
GRENADA BASIN
TOBAGO BASIN
BARBADOS RIDGE
GRENADA

World locator map:
North America
Area Enlarged
South America
ATLANTIC OCEAN

GULF OF MEXICO & THE CARIBBEAN SEA

The Atlantic Ocean encompasses 31,546,630 square miles (81,705,396 sq km), with a maximum depth of 28,232 feet (8,605 m) in the Puerto Rico Trench. The Mid-Atlantic Ridge, Earth's longest mountain range, is as long as the Andes, Rocky Mountains, and Himalaya combined. The Caribbean Sea and Gulf of Mexico give rise to hurricanes, some of Earth's most spectacular storms. Three nations border the Gulf, and 27 countries are within or border the Caribbean, a region heavily used by people for thousands of years for transportation, for extraction of fish and other ocean wildlife, and for prodigious quantities of oil and gas since the middle of the 20th century. Conservation efforts are under way to protect at least 20 percent of the Caribbean by 2020.

THE
OCEAN
IS
ALIVE

~

The bottom-line answer
to the question about why
biodiversity matters is fairly simple:
The rest of the living world
can get along without us,
but we can't get along
without them.

—SYLVIA A. EARLE

IN EVERY DROP, A MICROCOSM

THE BAHAMAS & THE SARGASSO SEA

Like liquid glass, moon jellies pulsed overhead through what seemed to be blue sky, air, and ocean merging across a mirror-calm sea, more than 200 miles from dry land. On a 1965 voyage from the Caribbean Sea to New York Harbor along the western edge of the Sargasso Sea, research vessel *Anton Bruun* paused to give scientists and crew a welcome open-sea swim where the bottom lay more than two miles down. Once overboard, beams of sunlight converged through silken water a few degrees cooler than my body temperature—my idea of a perfect swimming pool.

We were not alone in the sea with the jellies. Large mats of golden-brown *Sargassum* weed drifted like upside-down forests, with most of the leafy fronds extending underwater exposing just the uppermost clusters of blades, stalks, and berrylike floats to the surface. Looking closely, I could see eyes looking back! Tiny shrimp, miniature swimming crabs, juvenile flying fish, small jacks, a filefish, and a seahorse, all with speckled shell or skin matching precisely the color and pattern of the seaweed.

Lacy bryozoans, spiraling worm tubes, white-shelled barnacles, sea hares, nudibranchs, flatworms, amphipods, copepods, isopods, hydroids, anemones, and even encrusting sponges adorned the golden blades. Later, on the deck of the ship in a tubful of seaweed, I discovered as many of the major divisions of animal life as there are on some continents. Of the 35 or so major categories of animals, most have some representation in the ocean; only about 15 occur on all of the terrestrial parts of Earth combined.

Young tunas, loggerhead and green sea turtles, and other transients are sheltered in *Sargassum* forests during adolescent years spent at sea, and somewhere in the depths below, eels that have lived for decades in freshwater lakes and streams in Europe and North America periodically gather to spawn, setting in motion a daunting challenge for their hatchlings. In the eat-and-be-eaten planktonic wilderness, baby eels survive by being transparent and agile, but how, after three years of oceanic existence they find their way to the streams of their parents' origin remains a mystery.

Within a quart jarful of what appears to be lifeless, transparent seawater, scientists have in recent years discovered millions of individual microbes in thousands of varieties, truly a micro-universe. In the Sargasso Sea, in an area much like the place where I swam in the open ocean, samples of water were analyzed in 1986 using a new technique that made possible the discovery of organisms so small that they had escaped capture with nets, filters, and other methods previously used. Thus did Sally Chisholm, a scientist with Massachusetts Institute of Technology, and a team of expert colleagues, discover the cyanobacteria *Prochlorococcus,* a kind of photosynthetic organism that is responsible for about 20 percent of the oxygen in the atmosphere—one in every five breaths you take.

The name *Prochlorococcus* is not familiar to every schoolchild, government official, or corporate executive, but this may change as knowledge of the large significance of these

BRYCE GROARK ~ BERMUDA, NORTH ATLANTIC OCEAN
Sylvia Earle submerges in an ocean rich with plankton in Bermuda, an island nation in the Sargasso Sea, location of the northernmost coral reefs in the world.

minuscule microbes becomes more widely appreciated. Chisholm and others have continued investigations on several varieties of these small but mighty organisms, their notable impact on planetary processes and their wide, global distribution. Not only do they do much of the heavy lifting in terms of producing oxygen, but also, by taking up carbon dioxide and water in the presence of sunlight and chlorophyll, they produce sugar—the critical base of the great ocean food webs.

Among the most efficient grazers on phytoplankton are certain exquisitely constructed copepods and other crustaceans, including numerous species that are smaller than the commas punctuating this page. Copepods and other zooplankton engage in continuous interactions known as "energy exchange" or "nutrient cycles" but are quite simply an ongoing saga of eating and being eaten. Food powers each animal's activities, but some always goes back as a source of

eagles, and owls also eat grazers (zebras, deer, rabbits, mice), which in turn eat photosynthesizers (grass, shrubs).

Top carnivores in the ocean, creatures such as sharks, tunas, swordfish, marlin, dolphins, orcas, and sperm whales, make meals of other carnivorous animals that in turn have eaten other carnivores. They could starve in an ocean filled with phytoplankton because they are not able to see, let alone capture or use, a feast of micro-greenery. Middlemen are required, typically fish or squid that have eaten smaller fish that in turn have eaten those able to feed on the small grazers.

To make a pound of year-old chicken or plant-eating carp, catfish, or tilapia takes about two pounds of plants; for a pound of yearling cow, about 20 pounds of grass or grain. It takes many tons of phytoplankton at the base of a food chain involving legions of squid and small, medium, and large fish to produce a ten-year-old bluefin tuna, the age when that fish might begin to reproduce.

Diving into the ocean is like diving into
the history of life on Earth.

—SYLVIA A. EARLE

nutrient—fertilizer, if you will—which in turn is used by microbes such as *Prochlorococcus.* In healthy, natural systems, there is no such thing as "waste."

Copepods tend to reproduce quickly and profusely, a necessary strategy for survival considering the odds of being eaten by larger animals that are swimming in the same open-sea realm where hiding places are scarce. It takes a great many meals of phytoplankton to make a copepod, and many copepods to power the jellyfish, larval sea stars, arrow worms, baby squid, filter-feeding shrimp, barnacles, and small fish that dine on them. This stage in an ocean food chain—carnivores eating grazing animals that dine on photosynthesizers—is roughly equivalent to top carnivores on the land. Lions, tigers, wolves,

Sturgeon, some sharks, and many deep-sea fish may live to be 100. Farmers find it cost-effective to raise fast-growing, plant-eating animals, not long-lived carnivores, for food. Fishermen, however, as modern wildlife hunters, do not have to account for the enormous ecosystem investment in every fish, lobster, shrimp, or clam extracted from the sea. Swimming in the sea, the accounting base of sea creatures is zero, and we fail to subtract assets from the ocean's balance sheet when they are removed.

A member of the family Scombridae, bluefins are among the 60 or so torpedo-shaped fish that include mackerels, bonitos, and half a dozen other kinds of tunas, all critically important links in ocean ecosystems as top predators and as producers of nutrients that are returned to the ocean that in turn power plankton. All release millions of eggs into the sea during seasonal spawning, but only a lucky few make it through the gauntlet of hungry mouths to become adults themselves. Fewer still of the great bluefin tuna attain a life span of 30 years or so, or their majestic 15-foot, 1,500-pound potential.

Engineers admire bluefins as masters of speed, wondering at their ability to rocket through the ocean as fast as a nuclear submarine on a diet of fish and squid. As the tuna's tail moves back and forth, small vortices are formed and captured, yielding propulsion with 97 percent efficiency—far greater than any man-made system. Physiologists wonder at the unusual nature of bluefins and certain other high-speed fish that enables them to maintain a body temperature above that of ambient seawater. Warm-blooded fish! Fish with the uncanny ability to sense and react to water chemistry, temperature, and movement! Fish that often move in V-shaped formation reminiscent of migrating birds, and like some birds, travel thousands of miles over established pathways with no road map other than what is embedded in their brains through millions of years of history. Fish with elaborate courtship sensitivities that transcend what most people think of when they think of *fish*.

Sushi lovers prize bluefins as the ultimate in dining pleasure, a tradition in Japan but a recently acquired taste of luxury elsewhere in the world. In 1991, during my tenure as the chief scientist of NOAA, the National Oceanic and Atmospheric Administration, a paper crossed my desk with the news that since 1970, 90 percent of the North Atlantic Ocean's bluefin tuna had been taken. Shocked, I asked if our objective was to exterminate bluefins because if so, we were doing a great job. Just 10 percent left to go!

Thereafter, referred to by some as the "Sturgeon General," I tried to understand how we might succeed with policies that would make possible an ocean that functions in our favor, providing oxygen, taking up carbon, and otherwise holding the planet steady, while enabling us to enjoy a growing habit of

BRYCE GROARK ~ SARGASSO SEA, NORTH ATLANTIC OCEAN
The Atlantic gyre that fosters floating forests of Sargassum, *observed here by Sylvia Earle, also accumulates large quantities of deadly plastic debris, a growing threat to ocean wildlife globally.*

converting sea life into seafood. Noting the rapidly declining populations of turtles, tunas, swordfish, marlin, sharks, halibut, cod, herring, capelin, menhaden, grouper, snapper, oysters, clams, and many other wild animals, one thing is clear: There are limits, already surpassed, as to how much can be taken from ancient ocean ecosystems without inflicting damage to the healthy functioning of the living ocean. Moreover, fully protecting places such as the Sargasso Sea through national and international agreements is critical not just for the life history of turtles, tunas, and eels, nor for the floating golden forests of *Sargassum* and their richly diverse miniature menagerie. If you like to breathe, you will care about places such as the Sargasso Sea, about *Prochlorococcus,* about food webs and nutrient cycles, about ocean chemistry and the tightly knit connections among sea creatures that enable them—and humankind—to prosper. We can choose: continued loss or enhanced protection for the natural systems that keep us alive.

Never did I realize till now
what the ocean was:
how grand and majestic,
how solitary, and boundless,
and beautiful and blue . . .

—HERMAN MELVILLE, *REDBURN*

FLIP NICKLIN ~ BAHAMAS, NORTH ATLANTIC OCEAN
*A social pod of Atlantic spotted dolphins communicates
with sight, sound, and frequent touching as they glide
through their liquid universe.*

MIKE THEISS ~ BAHAMAS, NORTH ATLANTIC OCEAN
(*PRECEDING PAGES*)
*Swift currents carve grooves and channels in shallow
sands near the "Tongue of the Ocean" in the Bahamas.*

JO MAHY ~ BAHAMAS, NORTH ATLANTIC OCEAN
Looking like a diminutive patch of pumpkins,
a colony of corals host a peppermint goby.

ALEX MUSTARD ~ SHETLAND ISLANDS, SCOTLAND, NORTH ATLANTIC OCEAN
Colonial lightbulb sea squirts fluoresce green when photographed under deep
blue light. Related to vertebrates, their free-swimming larvae resemble tiny fish.

When we try to pick out anything by itself,
we find it hitched to everything else
in the universe.

—JOHN MUIR, *MY FIRST SUMMER IN THE SIERRA*

The last fallen mahogany
would lie perceptibly on
the landscape, and
the last black rhino would be
obvious in its loneliness,
but a marine species may
disappear beneath the waves
unobserved and the sea would
seem to roll on
the same as always.

—G. CARLETON RAY,
"ECOLOGICAL DIVERSITY IN COASTAL ZONES AND OCEANS"

BRIAN SKERRY ~ EAST COAST OF FLORIDA,
NORTH ATLANTIC OCEAN
*A mother right whale gets a playful bump from her
new calf along the Florida coast. The calf represents
hope for the future of a relentlessly hunted species,
now numbering about 300 individuals.*

My search for mind,
body and soul balance
brims at the bottom
of the ocean, for once
you arrive there,
the only other place
you can go is up.

—JASON MRAZ, MUSICIAN

JAD DAVENPORT ~ ANDROS ISLAND, BAHAMAS, NORTH ATLANTIC OCEAN
A valued natural member of Indo-Pacific reef communities, this elegant Pterois lionfish species has recently invaded and rapidly spread through the predator-depleted western Atlantic Ocean from New England to Brazil.

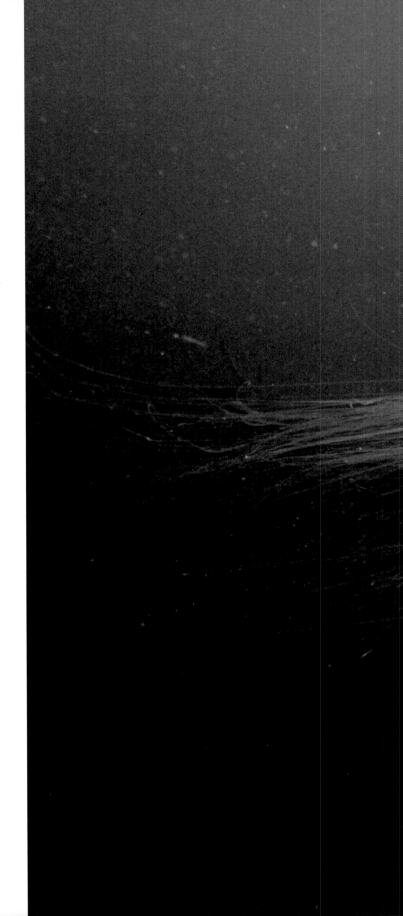

Ask me . . . where I'd most
like to be, and the answer is
always the same: in the water,
gliding weightless, slicing a
silent trail through whatever
patch of blue I can find.

—LYNN SHERR, *SWIM: WHY WE LOVE THE WATER*

SCOTT LESLIE ~ NOVA SCOTIA,
NORTH ATLANTIC OCEAN
*The body of the lion's mane jellyfish, one
of the largest known species of jellyfish,
can reach more than seven feet in diameter.*

When we take fish
from the ocean,
we are not farmers . . . ;
rather we are hunters . . . ,
more efficient and
more voracious than
any other inhabitant
of the marine food web.

—DANIEL PAULEY & JAY MACLEAN,
IN A PERFECT OCEAN

BRIAN SKERRY ~ ATLANTIC OCEAN
Caged bluefin tuna captured as juveniles are
being fattened for the sushi market, ending
their potential to become legendary ocean
giants that can live for decades.

Having wandered some distance
among gloomy rocks,
I came to the entrance of a great cavern . . .
Two contrary emotions arose in me,
fear and desire—fear of the threatening
dark cavern, desire to see whether there were
any marvelous things in it.

—LEONARDO DA VINCI, RENAISSANCE POLYMATH

ALEX MUSTARD ~ BAHAMAS,
NORTH ATLANTIC OCEAN
*Colorful sponges brighten the darkness of
Thunderball Cave near Staniel Cay in Exuma.*

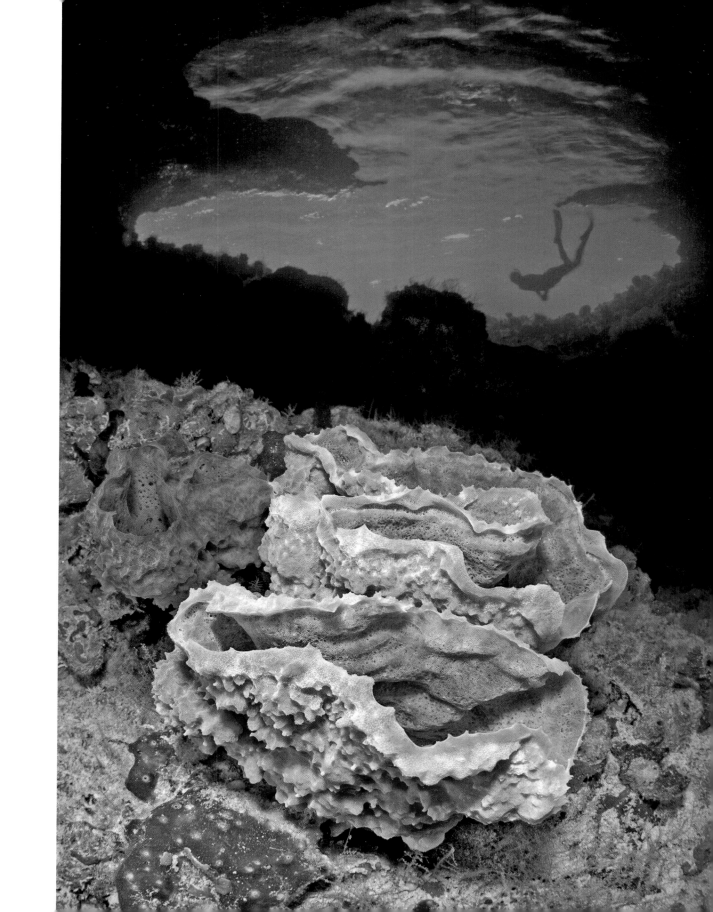

If we fail
to take care
of the ocean,
nothing else
matters.

—ADDISON FISCHER,
CONSERVATIONIST & BUSINESSMAN

ALEX MUSTARD ~ BAHAMAS,
NORTH ATLANTIC OCEAN
*Striped pilot fish accompany an oceanic
whitetip shark in the tropical western Atlantic
Ocean, off Cat Island.*

ALEX MUSTARD ~ SHETLAND ISLANDS, SCOTLAND,
NORTH ATLANTIC OCEAN
*A kelp frond hosts this tiny translucent nudibranch,
an ocean-dwelling cousin of snails and slugs.*

ALEX MUSTARD ~ NORWAY,
NORTH ATLANTIC OCEAN
*A diminutive beauty, this red-gilled nudibranch
vacuums sustenance from a blade of kelp.*

Exploring is
curiosity acted upon.

—DON WALSH, OCEAN ELDER

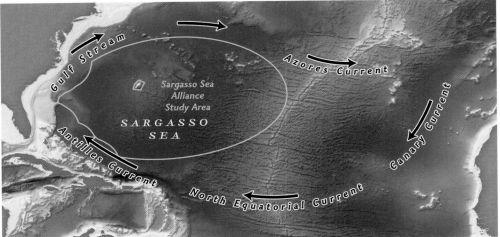

SARGASSO SEA

Liquid boundaries define the Sargasso Sea within the North Atlantic Ocean's subtropical gyre, an area largely within the High Seas. In 2014, representatives of various countries convened in Bermuda, and five nations signed the Hamilton Declaration, an expression of intent to work together to provide protective stewardship for the region and its unique "floating golden rain forest" of Sargassum seaweeds that generate oxygen, take up carbon dioxide, and provide shelter and sustenance for migratory fish, turtles, and whales, as well as permanent residence for many unique species.

EXCLUSIVE ECONOMIC ZONES

The concept for exclusive economic zones for nations bordering the ocean was formally adopted at the third United Nations Conference on the Law of the Sea in 1982, thereby greatly extending national jurisdiction over much of the ocean. About half of the world—64 percent of the ocean—lies beyond those boundaries and is considered international waters, a "global commons." Subject to various regional and international agreements, including the Law of the Sea, governance of the High Seas is currently under review with particular reference to transportation, fisheries policies, mining, and designation of protected areas.

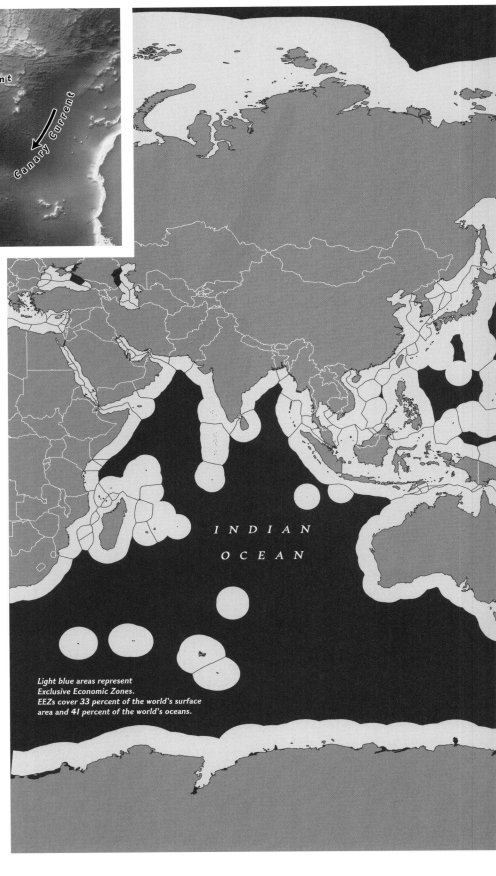

INDIAN OCEAN

Light blue areas represent Exclusive Economic Zones. EEZs cover 33 percent of the world's surface area and 41 percent of the world's oceans.

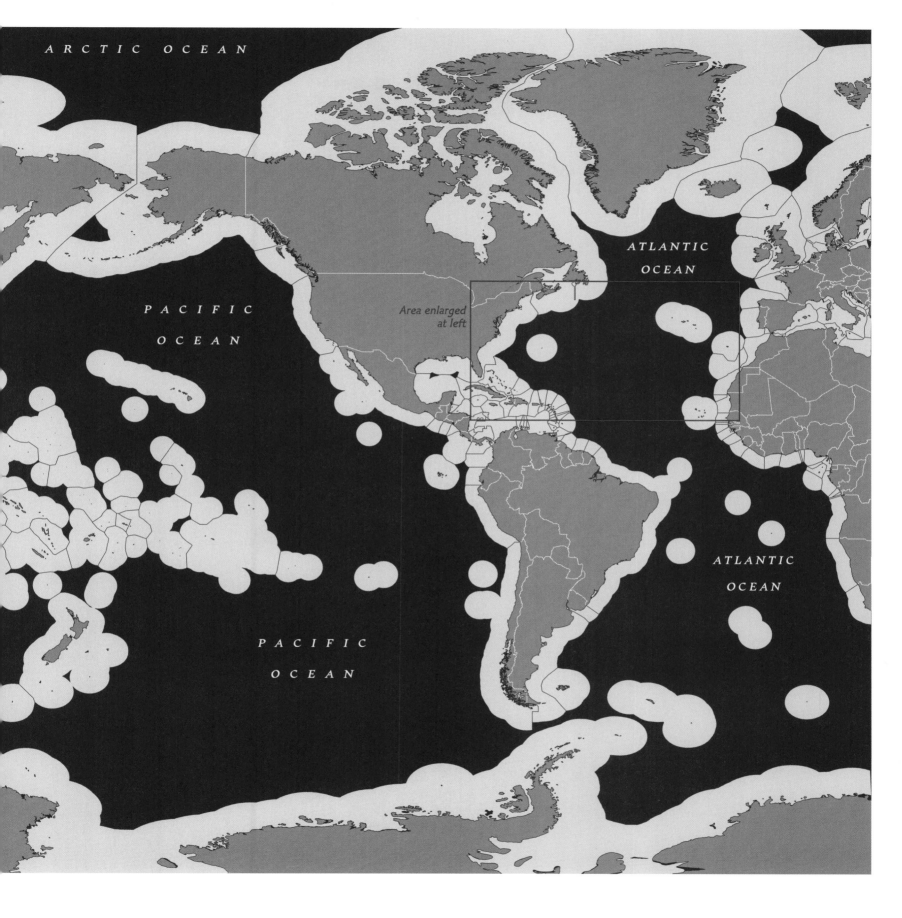

ARCTIC OCEAN

ATLANTIC
OCEAN

PACIFIC

OCEAN

Area enlarged
at left

ATLANTIC

OCEAN

PACIFIC

OCEAN

DEEP
FRONTIERS

~

More has been discovered about the nature of the ocean in the past century than during all preceding history.

—SYLVIA A. EARLE

THE GREATEST ERA OF EXPLORATION

THE INDIAN OCEAN

Sylvia Sails Away With 70 Men," read the 1964 headline in the *Mombasa Daily Times,* "but she expects no problems." As the only woman—and only botanist—aboard the U.S. research ship *Anton Bruun,* I was delighted to have 70 helpful male colleagues for a six-week National Science Foundation–sponsored International Indian Ocean Expedition, the last in a series of cruises during two years of unprecedented effort to understand the nature of one of the least known parts of the planet. The only *problems* I encountered were those that all shipboard scientists face as they try to understand the nature of the complex, living ocean using conventional instruments.

I imagined trying to understand cloud-shrouded New York City—from an aircraft flying high above towing a net through the streets, snaring a taxi, a few pedestrians, some bushes, a piece of a building, shards of glass, a dog or pigeon or two. From high in the sky, what could I discover about human society, our music, art, or sense of humor, or the connections that make our civilization function? What might Jane Goodall know of the habits of her beloved chimpanzees from a helicopter flying a mile above the forest? From the deck of a rolling ship, what could be learned about life in the sea miles below?

Every time a net came back on deck after being towed for an hour or so, scientists would crowd around like children pouncing on presents at a party. Each captured creature was like a piece of an enormous puzzle, with few clues about how one thing related to another. But aboard the *Anton Bruun* were some new tools for ocean exploration—masks, fins, scuba gear—that provided an exciting new way to observe creatures on their own terms and see how systems function by immersing ourselves directly into the sea.

Exploring the Comoro Islands with scuba in 1964 not only put me in contact with fish who had never seen a primate before, but also provided the tantalizing prospect of encountering a creature famously known as "Old Four-legs," the coelacanth, *Latimeria chalumnae.* Fossil coelacanths are known from 400 million years ago, and the animals were thought to have become extinct along with the dinosaurs—until fishermen discovered one near Port Elizabeth, South Africa, in 1938. A few years later, one was captured hundreds of miles to the north in the Comoro Islands—right where we were exploring.

No coelacanths swam into view during my dives in 1964, but I was entranced by face-to-face encounters with placid green sea turtles, curious blue parrotfish, elegant green wrasses, and intricate communities of sponges, brittle stars, corals, and other residents of pristine reefs. Jumbled and crushed in a net, I might find as many species as I could by diving, but it would not be possible to know how they are arranged, move, interact, and otherwise naturally behave. A dusty, dried, hollow-eyed specimen of *Latimeria* on display at the small airport on Grand Comoro Island provided tangible evidence that we were in the neighborhood of these age-old relics, but observations of living coelacanths in their deepwater lairs were not made until many years later when a small

KIP EVANS ~ ATLANTIC OCEAN
Submersibles such as this, piloted by Sylvia Earle, are "so simple to drive, even a scientist can do it." They were used during the National Geographic/NOAA Goldman Foundation Sustainable Seas Expeditions.

submarine transported scientists into subsea canyons nearby.

Divers breathing compressed air are limited by physiological constraints to about 200 feet, but submarines, like spacecraft, provide safe havens of air at sea-level pressure in otherwise hostile environments. Getting into a submarine is like taking a taxi far below where divers can venture, with the added advantage of having more time and dry quarters for snacks, instruments, and communication with those on the surface. Some—such as the first research sub I had a chance to use, the Perry-Link *Deep Diver*—have two compartments,

After completing a series of dives in Hawaii in 1979 using *Jim,* a one-person submersible featuring metal arms and legs, I was inspired to co-found a company to build a new generation of small subs, starting with a 3,200-foot system we named *Deep Rover* and later, developing hundreds of small underwater robots. Like many California start-up companies, the enterprise began in my garage, and soon 15 engineers were ensconced with their drawing boards, calculators, and piles of technical papers on the porch, in the living room, in the family room, and sometimes in the backyard—a place excavated to

Most of the ocean has yet to be seen let alone explored or mapped with accuracy comparable to Mars or the far side of the moon.

—SYLVIA A. EARLE

one for a pilot and observer, and the other for divers who can pressurize their enclosed space to be the same as that of the surrounding ocean, lift a hatch in the floor, and swim out for brief open-ocean explorations using mixes of gases that extend the range below air-diving depths.

I was expecting my daughter, Gale, in 1968 when I stepped from *Deep Diver* into a cool blue ocean to experience this approach to research. Years later, Gale explains her love of the ocean as inevitable. "After all," she says. "I was diving before I was born." Each of my three children has shared with me—as witness and active participant—in what has turned out to be the most extraordinary time of ocean exploration in human history. Prior to the 1960s, no one had been to the ocean's greatest depth, microbes in the ocean were thought to be rare, the 40,000-mile-long Mid-Ocean Ridge that extends like a giant backbone down the major ocean basins had yet to be defined, together with plate tectonics as the basis for the movement of continents and the existence of hydrothermal vents and their associated gardens of life.

accommodate a test-tank that on hot summer days doubled as a swimming pool.

The 1980s marked an era of exceptional activity in the development of both robotic and manned systems for underwater exploration for science—and in support of salvage, laying cable, mining, underwater construction, and rapidly growing offshore oil and gas industries. Small robotic vehicles tethered to surface controls that scientists could drive while looking at images conveyed to television monitors began to complement heavy-duty, car-size industrial systems. While proving their worth as avatars of their surface-based pilots, I soon came to realize that even highly sophisticated remotely operated systems cannot fully replace the freedom and finesse of a diver, or the ineffable experience of personally "being there" in a submarine.

In *Deep Rover,* and later in dozens of dives in the Canadian-built system called *Deep Worker,* I would often find a place to settle down on the bottom, shut off the lights, and simply watch the flow of small organisms drifting by, some emitting a constant spark of blue light, others flashing a burst of

Off the coast of the African country Mauritania, a trawler hauls in massive amounts of fish, including many that are discarded as "bycatch."

brilliance as they collided with the sub's clear dome or sphere. Once, in 1,300 feet of water, I glimpsed a shadowy form beyond the range of my cameras and turned the sub for a better view of what turned out to be an octopus as large as I am. With a remotely operated system, I would have surely missed seeing what I discovered was a mother mollusk, embracing a mass of embryonic octopuses, and would definitely not have been able to execute the slow-motion dance that followed. Little subs are nimble and easy to drive, allowing me to spend an hour while slowly ascending to 1,000 feet, with the octopus moving away until I paused, returning when I stopped, seemingly as curious about the sub and its contents as I was about her. She was inches from my face when the sub's battery power ran low and I had to make my way back to the surface.

New technologies have not only greatly increased knowledge of the ocean and the creatures who live there; they have also been used to find, capture, freeze, transport, and market previously inaccessible ocean wildlife on an unprecedented scale. From thousands of feet below the surface, long-lived grenadiers or "rat tails" have been caught and sold as "hoki" in fast food restaurants. Deep-dwelling Patagonian toothfish marketed as "Chilean sea bass" have become a favorite gourmet treat. Fish sold as "orange roughy," known to scientists as "slime heads," take 30 years to mature and may have lived more than a century when captured for human consumption. The populations of these and numerous other sea creatures that have never experienced humans as predators until recent years are swiftly collapsing. Exploitation has raced far ahead of exploration as places that are barely mapped, let alone researched, are stripped of life.

But attitudes are beginning to change. In 1981, I participated in a gathering of leaders from 16 Indian Ocean countries to establish the Indian Ocean as a protected area for whales. Since 1986, most commercial whaling has stopped, and marine mammals generally are becoming respected more for their value alive than when turned into pounds of meat and barrels of oil. Some seabirds were exploited to near extinction for meat, feathers, and eggs until they, too, became recognized for their greater value alive. Although too late to save the great auk, last seen alive in 1852, or the Steller's sea cow, eliminated in 1768 or the Caribbean monk seal, lost in 1952, various forms of protection have helped restore many from the edge of oblivion.

Similarly, fish, squid, lobsters, oysters, crabs, and other ocean wildlife have been relentlessly exploited using wondrous new technologies developed in recent years—some to a level where recovery may not be possible. But for these creatures, too, attitudes are changing, and the importance of intact ocean ecosystems is gaining favor. In 2010, the United Kingdom designated a fully protected marine reserve around the 55 islands of the Chagos Archipelago in the Indian Ocean, an area more than twice as large as the United Kingdom itself. The action is part of a growing trend globally to safeguard parts of the ocean for values that transcend extractive uses coincident with awareness that the most important thing we take from the ocean is our *existence*. The greatest era of exploration—and care—of the ocean has just begun.

How do you get to the
great depths? . . . How do you
return to the surface
of the ocean? And how do
you maintain yourselves
in the requisite medium?
Am I asking too much?

—PROFESSOR PIERRE ARONNAX IN *TWENTY THOUSAND
LEAGUES UNDER THE SEA* BY JULES VERNE

CESARE NALDI ~ ANDAMAN SEA
*Equipped with a natural snorkel, a 60-year-old
elephant takes a dip in the warm waters of the
Andaman Sea, off the eastern coast of India.*

**JAMES STANFIELD ~ REPUBLIC OF THE MALDIVES,
INDIAN OCEAN**
(PRECEDING PAGES)
*Barely emergent from the reach of high tide and
under threat of being engulfed by rising sea level,
the islands of Maldives make up a small country
in terms of land, but have jurisdiction over nearly
400,000 square miles of the Indian Ocean.*

One touch of nature
makes the whole world kin.

—WILLIAM SHAKESPEARE,
TROILUS AND CRESSIDA

NORBERT WU ~ THE SEYCHELLES, INDIAN OCEAN
*A lone bigeye fish accompanies doubletooth
soldierfish, which rest by day in coral caves and
emerge to feed at night.*

NORBERT WU ~ THE SEYCHELLES, INDIAN OCEAN
A candy cane sea star rests atop Acropora coral in the Seychelles.

NORBERT WU ~ THE SEYCHELLES, INDIAN OCEAN
A pair of spotted porcelain crabs perch on their doorstep, the slippery edge of a giant anemone.

It's exciting—humbling, really—to be the first
to see a fish or other marine critter
that no one else has ever seen before.

—RICHARD PYLE, TED TALK: "A DIVE INTO THE REEF'S TWILIGHT ZONE"

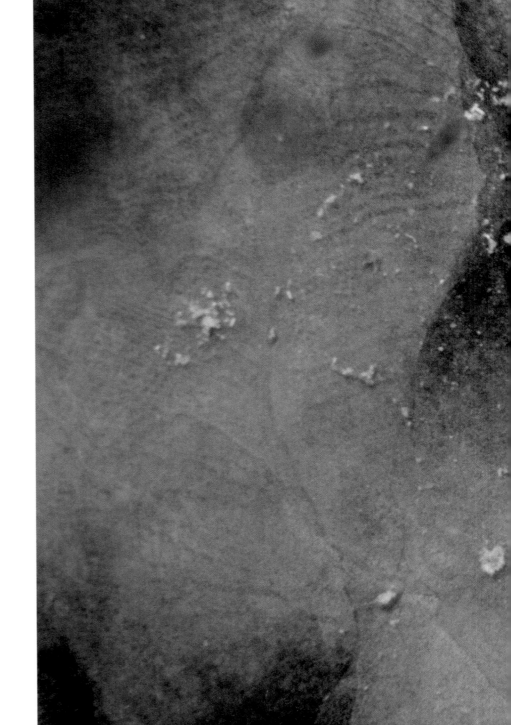

When we embark on the
great ocean of discovery,
the horizon of the unknown
advances with us and
surrounds us wherever
we go. The more we know,
the greater we find
is our ignorance.

—GARDINER G. HUBBARD,
NATIONAL GEOGRAPHIC, VOL. 1, 1888

PAUL SUTHERLAND ~ REPUBLIC OF THE MALDIVES,
INDIAN OCEAN
*A parrotfish stares into the lens of a camera on
a reef in the Maldives.*

It seems the ocean's
chain of life is actually
a fragile, silken web.
If just one strand is removed,
the whole thing unravels.
And it may never be
whole again.

—MARLA COHEN, ENVIRONMENTAL WRITER

THOMAS PESCHAK ~ REPUBLIC OF THE MALDIVES,
INDIAN OCEAN
*Gentle giants, manta rays move with colossal
grace while dining on microscopic zooplankton
in the Maldives.*

We are tied to the ocean.
And when we go back to the
sea, whether it is to sail or to
watch it, we are going back
from whence we came.

—PRESIDENT JOHN F. KENNEDY

LAURENT BALLESTA ~ SOUTH AFRICA,
INDIAN OCEAN
Relicts from ancient seas, coelacanth fish were
known only from fossils until 1938, when one
was captured by fishermen not far from where
divers recently encountered this individual living
in the deep waters of Sodwana Bay.

Who knows what admirable
virtue of fishes may be below
low-water mark, bearing up
against a hard destiny,
not admired by that
fellow creature who alone can
appreciate it! Who hears
the fishes when they cry?

—HENRY DAVID THOREAU,
A WEEK ON THE CONCORD AND MERRIMACK RIVERS

JAD DAVENPORT ~ MOZAMBIQUE,
INDIAN OCEAN
*Clusters of anthia fish gather on a reef in the
Quirimbas Archipelago off the northeastern
coast of Mozambique.*

The bay

is where

thousands of organisms thrive

us

we look up

and see you

looking down at us

trying to understand us,

how we live, how we manage

to hold the seas up

keeping them in balance

and

you try to learn

how to keep the world in balance too

and you try

and try

and maybe someday

you will learn

to hold the world up

the right way

—EVAN HANKE, STUDENT

LUIS MIGUEL CORTES ~ INDIAN OCEAN
The azure waters of the Indian Ocean are the
open-ocean home for this green sea turtle.

Those who contemplate
the beauty of the earth find
reserves of strength that will
endure as long as life lasts.

—RACHEL CARSON,
THE SENSE OF WONDER: STORIES OF WORK

JAD DAVENPORT ~ REPUBLIC OF THE MALDIVES,
INDIAN OCEAN
*Schooling longfin bannerfish and fusiliers
patrol a rocky section of Shaviyani Atoll,
in the Maldives.*

DAVID SHALE ~ INDIAN RIDGE, INDIAN OCEAN
A deepwater seamount near Dragon Vent on the southwest Indian Ridge is home to this bristly armed brittle star.

DAVID SHALE ~ INDIAN RIDGE, INDIAN OCEAN
Living in a glass house: A tiny ghost shrimp resides within the lacy spicules of a deepwater glass sponge from Indian Ridge.

What is a scientist after all?
It is a curious man looking through a keyhole,
the keyhole of nature,
trying to know what's going on.

—JACQUES-YVES COUSTEAU, OCEANOGRAPHER

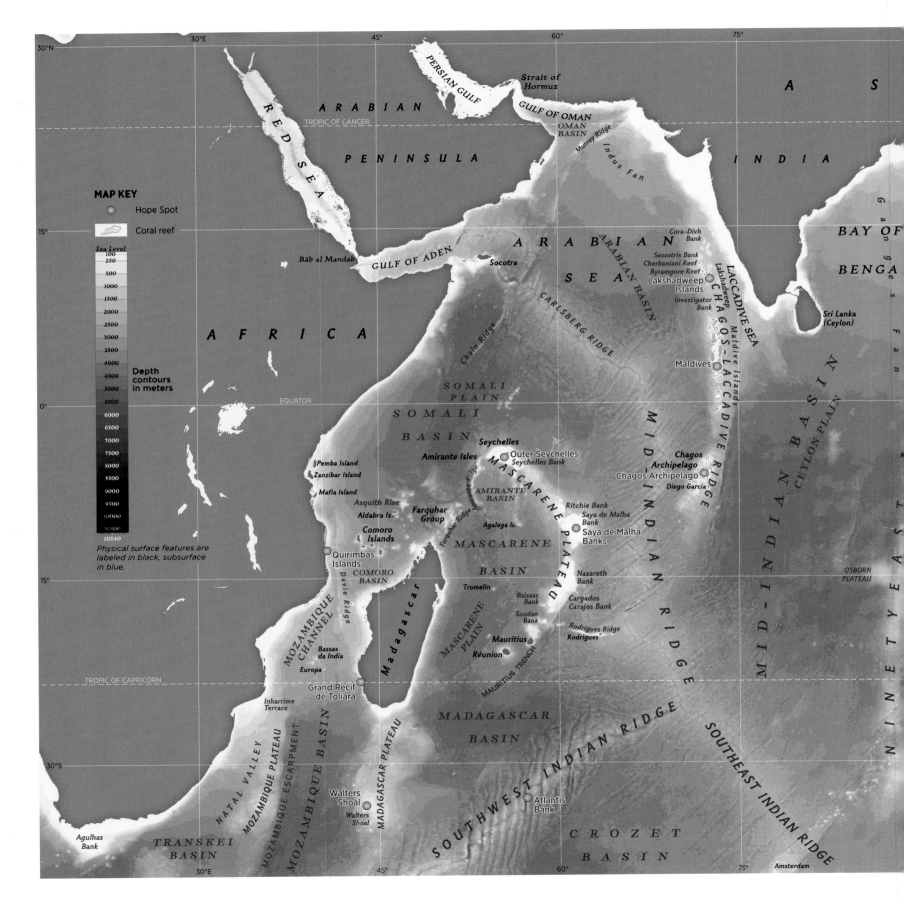

MAP KEY

◎ Hope Spot

Coral reef

Sea Level
100
250
500
1000
1500
2000
2500
3000
3500
4000
4500
5000
5500
6000
6500
7000
7500
8000
8500
9000
9500
10000
10500
10540

Depth contours in meters

Physical surface features are labeled in black, subsurface in blue.

30°N

30°E   45°   60°   75°

PERSIAN GULF

Strait of Hormuz

ARABIAN

GULF OF OMAN

A   S

TROPIC OF CANCER

RED SEA

PENINSULA

OMAN BASIN

Murrey Ridge

Indus Fan

INDIA

Bāb al Mandab

GULF OF ADEN

Socotra

ARABIAN

SEA

ARABIAN BASIN

CARLSBERG RIDGE

Cora-Divh Bank

Sesostris Bank
Cherbaniani Reef
Byramgore Reef

Lakshadweep Islands

Lakshadweep

LACCADIVE SEA

CHAGOS-LACCADIVE RIDGE

Maldive Islands

BAY OF

BENGA

15°

AFRICA

Investigator Bank

Sri Lanka (Ceylon)

Ganges Fan

Chain Ridge

SOMALI PLAIN

SOMALI

BASIN

Maldives

MID-INDIAN BASIN

CEYLON PLAIN

EQUATOR

0°

Seychelles

Amirante Isles

Outer Seychelles
Seychelles Bank

MASCARENE PLATEAU

Chagos Archipelago

Chagos Archipelago

MID-INDIAN RIDGE

NINETYEAST

Pemba Island
Zanzibar Island

AMIRANTE BASIN

Diego Garcia

Mafia Island

Asquith Rise

Aldabra Is.

Farquhar Group

Farquhar Ridge

Amirante Ridge

Agalega Is.

Ritchie Bank
Saya de Malha Bank

Saya de Malha Banks

Comoro Islands

MASCARENE

OSBORN PLATEAU

Quirimbas Islands

COMORO BASIN

BASIN

Tromelin

Nazareth Bank

15°

Davie Ridge

MOZAMBIQUE CHANNEL

Madagascar

MASCARENE PLAIN

Baissac Bank

Soudan Bank

Cargados Carajos Bank

Rodrigues Ridge

Rodrigues

Bassas da India

Europa

Mauritius

Réunion

MAURITIUS TRENCH

Inharrime Terrace

Grand Recif de Toliara

MADAGASCAR

BASIN

TROPIC OF CAPRICORN

30°S

NATAL VALLEY

MOZAMBIQUE PLATEAU

MOZAMBIQUE ESCARPMENT

MOZAMBIQUE BASIN

MADAGASCAR PLATEAU

Walters Shoal

Walters Shoal

Atlantis Bank

SOUTHWEST INDIAN RIDGE

SOUTHEAST INDIAN RIDGE

CROZET

BASIN

Agulhas Bank

TRANSKEI BASIN

Amsterdam

30°E   45°   60°   75°

INDIAN OCEAN

The Indian Ocean covers 26,050,135 square miles (67,469,539 sq km) of Earth with an average depth of 12,785 feet (3,897 m) and a maximum depth of 23,276 feet (7,125 m) in the Java Trench south of Indonesia's arc of islands. In 2010, the government of the United Kingdom established the largest marine no-take reserve in the world, more than 247,105 square miles (640,000 sq km) around the Chagos Archipelago, encompassing an area more than twice as big as the terrestrial part of the country. From abyssal deep water to pristine coral reefs, the reserve protects a vital cross section of ocean life, from migratory tunas, turtles, and whales to small creatures that are unique to the region.

CARING
FOR
THE
OCEAN

You may never see the ocean,
you may never touch the ocean,
but the ocean touches you—
with every breath you take,
every drop of water you drink.

—SYLVIA A. EARLE

SAFE HAVENS IN THE SEA

THE GREAT BARRIER REEF & CORAL SEA

Watch out for the sharks! There're man-eaters down there! I heard this often when I began diving in the 1950s. Long before author Peter Benchley struck terror in the minds of many about sharks with his blockbuster novel and film *Jaws,* people imagined sharks to be among the most fearsome and deadly creatures on Earth. It occurred to me that as a woman, I did not qualify for the attention of *man*-eating animals, but mindful of their carnivorous habits, the sight of the distinctive, sleek form of a shark lifted the fine hairs on the back of my neck during frequent underwater encounters.

After all, many sharks are large predators responsible for feeding themselves, and everything I have ever done, all that I care about, all of my hopes and dreams would not matter to hungry sharks for whom I might be just another tasty meal. Being eaten alive stirs deeply primal feelings, a sobering thought that came into sharp focus in 1976 while diving near Marion Reef, an area of small emergent coral reefs anchored in the deep water of the Coral Sea, more than 100 miles off-shore from Australia's eastern coast.

While standing on a patch of sand with my longtime friend and filmmaker, Al Giddings, in 70 feet of water, more than 50 gray reef sharks wheeled slowly around us, a silver-gray halo of fins, teeth, and torpedo-shaped bodies, each about twice my size. "In their eyes, I am just a piece of meat," I thought. Many people view fish primarily—or solely—as food. Why shouldn't fish similarly regard us?

But on this and thousands of other dives, I was astonished to find that with rare exceptions, people are not on the menu of sharks or other sea creatures who have the option of tearing us apart, but do not do so. So often did sharks pass me by as something not worth tasting that I was not only relieved but also slightly offended. Rejected by sharks! Like legions of others who began diving in the 1950s and millions who have since taken the plunge, I began to experience the reality about sharks. Nothing in their 300 million years of history prepared them to regard us as food. Moreover, nothing has prepared them to be able to defend themselves against humans who now kill hundreds of millions of them every year for shark steaks, shark fin soup, for their cartilage as an ingredient in certain pharmaceuticals, as incidental bycatch while fishing for other marine life, and as the target of killing for sport.

In 1980, designated as a "Year of the Ocean" by the U.S. government, a Department of Commerce campaign was initiated to encourage the catching and marketing of what they referred to as "underutilized species." Sharks were primary targets owing to their apparent abundance, combined with a lack of consumer interest other than in Asia.

Within 30 years, the success of creative marketing, generous fishing subsidies, and the lure of "free goods" with an accounting base of zero, coupled with use of wartime technologies applied to enhanced fishing capacity—sonars, satellite navigation, better weather forecasts, more reliable ships, improved refrigeration, freezing, processing and transporting methods, and nets and lines made of durable, low-cost

BRYCE GROARK ~ CORAL SEA
Sylvia Earle returns in 2013 to reefs in the Coral Sea, which she first viewed in 1976.

synthetics—made possible the swift depletion of 90 percent of the sharks and many other wild fish species globally. The greatest ocean predators had more than met their match. But during the same three decades, attitudes began to shift. Swift losses of sharks and other wild fish occurred in parallel with a growing sense of concern that an ocean without large predators would result in serious disruptions to ocean ecosystems, which in turn would have harmful consequences to the nature of the ocean— and eventually to global processes that work in our favor.

As sharks *decreased,* interest *increased* in protecting them for their beauty, their ecological significance, and for what proved to be their star power for attracting tourists. When a young great white shark was rescued from a fisherman's net near Monterey, California, and exhibited at the Monterey Bay Aquarium in the 1990s, thousands of visitors flocked to see her. The film *Finding Nemo* portrayed sharks as allies, not enemies of the bite-size lead characters. Conservation organizations

conference in New York City that henceforth, *all* commercial fishing would cease throughout its EEZ, thus providing sanctuary status for all creatures, both small—lobsters, shrimp, sea cucumbers, sea stars, parrotfish, butterfly fish, and the treasured nautilus—and great—tuna, grouper, snapper, swordfish, barracuda, manta ray, and the intricately patterned Napoleon wrasse.

Sharks thus have helped lead a sea change in the way people regard ocean wildlife. Policies and habits that formed when the ocean was largely intact and the number of people was far fewer than at present still drive large-scale extraction of ocean wildlife as commodities. However, three issues now drive new ways of thinking about the value of the living ocean: economic reality; ecological reality, with humans fully engaged as a part of the natural world; and an ethical awakening of the responsibility we now have to future generations—both of people and of life in the ocean.

> We must protect the ocean as if our lives depend on it —because they do.
>
> —SYLVIA A. EARLE

teamed with scientists to establish evidence of the critical importance of protecting sharks, and tourist resorts calculated that an individual shark alive could yield decades of revenues worth more than a million U.S. dollars. A dead shark might net fishermen a few hundred dollars—once. No one yet has figured out how to make a shark from scratch.

Palau, a Pacific Island nation of 21,000 people, became the first country to protect sharks within its entire 230,000-square-mile exclusive economic zone (EEZ), noting with businesslike rationale that sharks alive are more valuable to their tourism-based economy than sharks dead. In 2014, I listened to Palau's President Tommy Remengesau announce at a United Nations

In the Cook Islands in 2012, I attended one of the annual meetings of 16 small island/big ocean nations in the company of Conservation International's president, Peter Seligmann, and CI's chief ocean scientist, Greg Stone, to witness the ways these nations are working together to consider best practices for their sometimes overlapping blue frontiers. The Cook Islands president, Henry Puna, expressed a resolve to be independent of fossil fuel imports by 2020 and to implement protection for at least half of its EEZ—about a million square miles of ocean. President Anatole Tong, leader of the low-lying nation, Kiribati, described how his country is working with CI to fund alternative revenues to current fishing leases that underpin the nation's economy. Together, these nations are viewing their part of the planet as an "oceanscape" with mutual benefits if managed with care, mindful that they have a choice: gaining short-term profits from large-scale fishing, mining, and destructive developments, or winning sustainable wealth from their ocean assets—using the ocean, but not using it *up*.

When established in 1975, the Great Barrier Reef Marine Park Authority marked the world's biggest, boldest, most notable ocean conservation initiative in the world. Actions in 2014 led by Australia Minister for the Environment Tony Burke would be even larger, bolder, not only for protection of "blue Australia," but also as a meaningful global precedent. As part of an evaluation of the entire coastline of the country over many years, 30 percent would be marked for protection. Half of the Coral Sea was designated for full protection, the rest for management of fishing and care for historically treasured sunken relics from World War II.

As part of making the film *Mission Blue,* I returned to Australia's Coral Sea in 2013 aboard the yacht *Ethereal* to revisit places I had first witnessed in the 1970s. I wondered if the sharks I had seen then—or their descendants—might still be thriving at Marion Reef or other pristine sites observed years before: Osprey, Holmes, and others. Holmes Reef was our first stop during the 2013 Coral Sea expedition. Clear water and the shadowy forms of reefs coming to within 66 feet of the surface looked promising as we slipped from warm air to a warm sea, more than 150 miles from the mainland. I tempered my expectations of what we would find, knowing that long-line fishing and other commercial operations had extracted large numbers of sharks and other open-ocean animals in the decades since my most recent visit. But I wasn't prepared for the reality of *no* sharks, and few large fish of any kind. Recent storms had apparently damaged some of the corals, and there were clusters of brilliant soft corals, sponges, and a few large anemones with their attendant clown fish.

BRIAN SKERRY ~ GULF OF CALIFORNIA, MEXICO
Guitarfish, manta rays, and other fish are tossed from a shrimp boat, typical of the collateral damage incurred when trawls are dragged across the seafloor. Bycatch typically far exceeds the volume of shrimp retained.

With protection, it is likely that this part of the Coral Sea will be returned to health, joining intact systems that remain and robust areas within the adjacent Great Barrier Reef. At a 2003 conference in Mexico where 150 representatives from 20 countries and 70 organizations representing government, industry, academia, and conservation had gathered to deliberate the future of the ocean, opening remarks by Graeme Kelleher set the tone. As founding director of the Great Barrier Reef Marine Park Authority, he had personally experienced the era of transition from looking at the ocean as a place of infinite resilience to one that requires infinite care. He noted the convergence of knowledge that has led to understanding the need to step back from uncontrolled exploitation of the ocean to understanding the limits—and exercising our power by managing ourselves responsibly. Quoting Shakespeare, he noted: "There is a time in the tides of men . . . which taken at the flood leads on to fortune . . . on such a full sea are we now afloat. And we must take the current when it serves, or lose our ventures." He added: "This is the time."

I look at the world
like it has two oceans.
The ocean around us,
and the ocean of sky above,
the two big Blue Oceans!

—DAVID DE ROTHSCHILD,
FOUNDER, ADVENTURE ECOLOGY

**DAVID DOUBILET ~ GREAT BARRIER REEF,
SOUTH PACIFIC OCEAN**
*A marine scientist dives above a garden of stony
corals of the Great Detached Reef, in the far
northern section of the Great Barrier Reef.*

**ANDREW WATSON ~ GREAT BARRIER REEF,
SOUTH PACIFIC OCEAN**
(PRECEDING PAGES)
*A valentine for the ocean, Heart Reef hosts
millions of small creatures at Hardy Reef, within
the Great Barrier Reef on Australia's eastern coast.*

Oceans are critical . . .
Those dark-blue waters
are perhaps the single
greatest natural treasure
on God's Earth.

—VICE PRESIDENT AL GORE

CHRIS NEWBERT ~ SOLOMON ISLANDS,
SOUTH PACIFIC OCEAN
*A whirlpool of blackfin barracuda spiral at a
depth of 70 feet in the Solomon Islands in the
South Pacific Ocean.*

FRED BAVENDAM ~ GREAT BARRIER REEF,
SOUTH PACIFIC OCEAN
Serpentine arms of a brittle star lace around a plump indigo sea star. Both are active hunters of small prey amid corals in the Great Barrier Reef.

MELISSA FIENE ~ GREAT BARRIER REEF,
SOUTH PACIFIC OCEAN
A mosaic jellyfish drifts 100 miles offshore from Cairns, Australia.

O Lord, how manifold are thy works! . . .
So is this great and wide sea, wherein are things
creeping innumerable, both small and great beasts.

—PSALMS 104:24-25

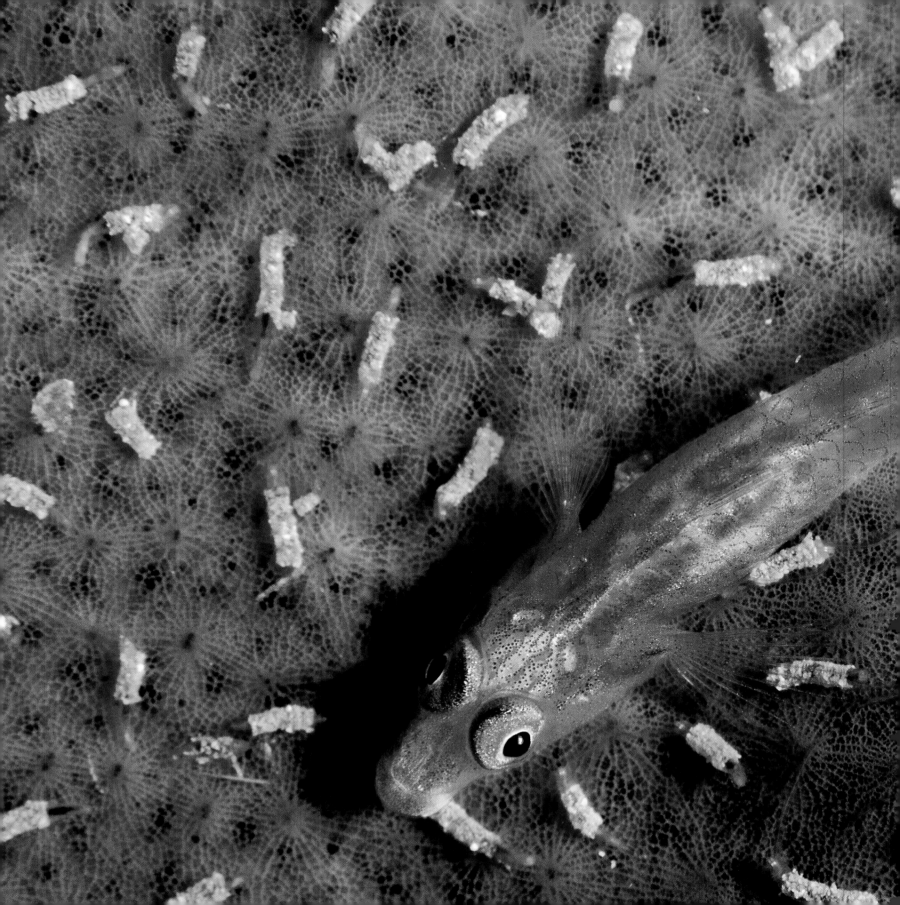

Penetrating so many secrets,
we cease to believe
in the unknowable.
But there it sits,
nevertheless,
calmly licking its chops.

—H. L. MENCKEN, *MINORITY REPORT*

JENNIFER HAYES ~ GREAT BARRIER REEF,
SOUTH PACIFIC OCEAN
*Home for this tiny goby is an elephant
ear sponge in Challenger Bay on the
Great Barrier Reef.*

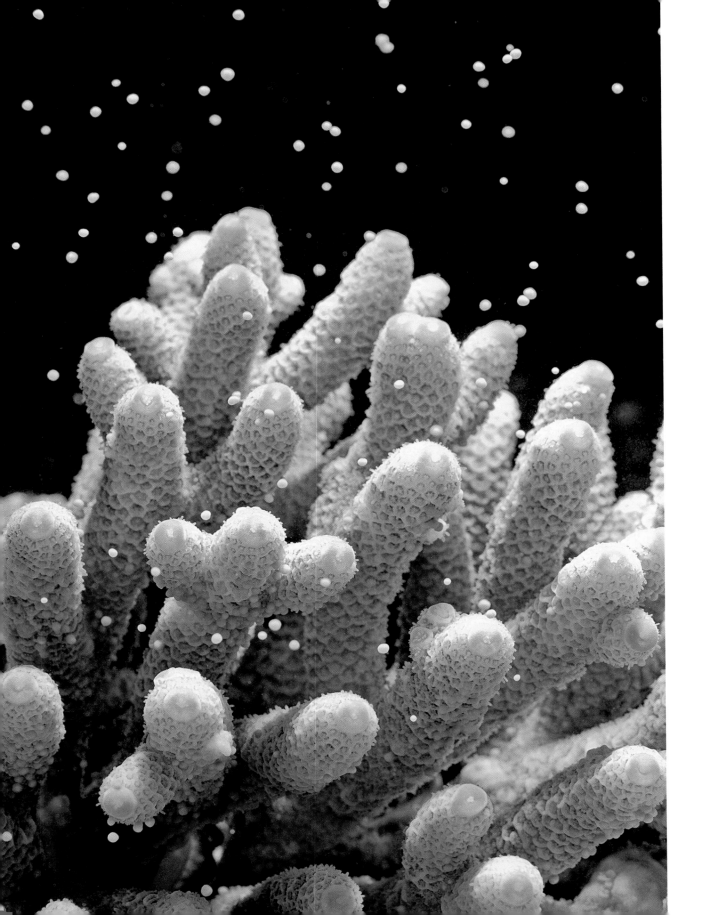

I will go back to the great sweet mother,
Mother and lover of men, the sea. . . .
O fair white mother, in days long past
Born without sister, born without brother,
Set free my soul as thy soul is free.

—ALGERNON CHARLES SWINBURNE, "THE TRIUMPH OF TIME"

DAVID DOUBILET ~ GREAT BARRIER REEF,
SOUTH PACIFIC OCEAN
Acropora millepora *coral releases egg and
sperm bundles simultaneously near Heron
Island, in the southern Great Barrier Reef.*

The sea can be so wondrously amazing! I love flying weightless in the water and seeing that gorgeous, strange and fascinating world under the sea.

—DARYL HANNAH, ACTRESS & ENVIRONMENTALIST

DAVID DOUBILET ~ GREAT BARRIER REEF, SOUTH PACIFIC OCEAN
Distinctively different patterns of stripes and dots mark the faces and flanks of each of these sweetlip fish in Challenger Bay, Great Barrier Reef.

CHRIS NEWBERT ~ SOLOMON ISLANDS,
SOUTH PACIFIC OCEAN
Forty feet under the sea near the Solomon Islands,
a feather duster worm spreads its soft array
of bristles.

FRED BAVENDAM ~ CORAL SEA,
SOUTH PACIFIC OCEAN
Most of the relatives of this chambered nautilus,
referred to as a living fossil, lived in ancient oceans
hundreds of million years ago.

Of all objects that I have ever seen,
there is none which affects my imagination
so much as the sea or ocean.

—JOSEPH ADDISON, *THE SPECTATOR*

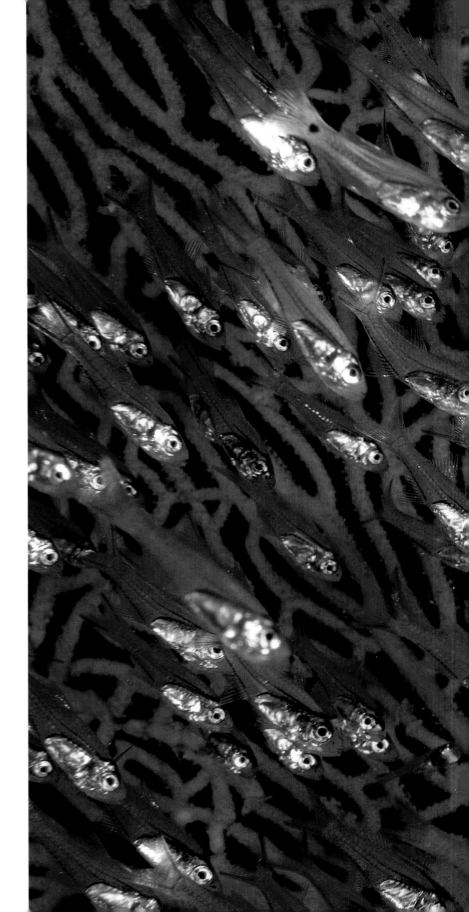

We have to find the right balance between what is good for each individual and what is good for all of us.

—ELLIOTT A. NORSE, FOUNDER & CHIEF SCIENTIST, MARINE CONSERVATION INSTITUTE

DAVID DOUBILET ~ GREAT BARRIER REEF, SOUTH PACIFIC OCEAN
Cardinalfish gleam against a gorgonian sea fan in Three Sisters Reef, part of Milln Reef.

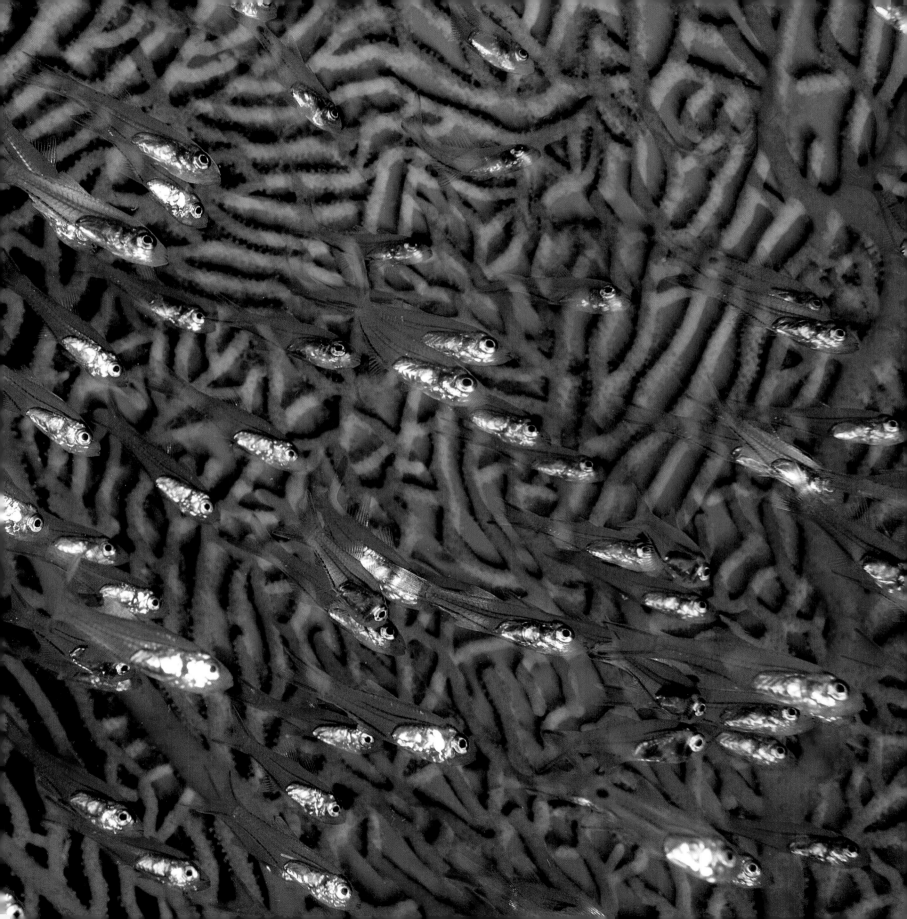

We shall not cease
from exploration,
And the end of all
our exploring
Will be to arrive where
we started
And know the place
for the first time.

—T. S. ELIOT, "LITTLE GIDDING"

DAVID DOUBILET ~ GREAT BARRIER REEF,
SOUTH PACIFIC OCEAN
Distinctive in appearance and in attitude,
humphead wrasses are beloved by divers for
their beauty and engaging curiosity.

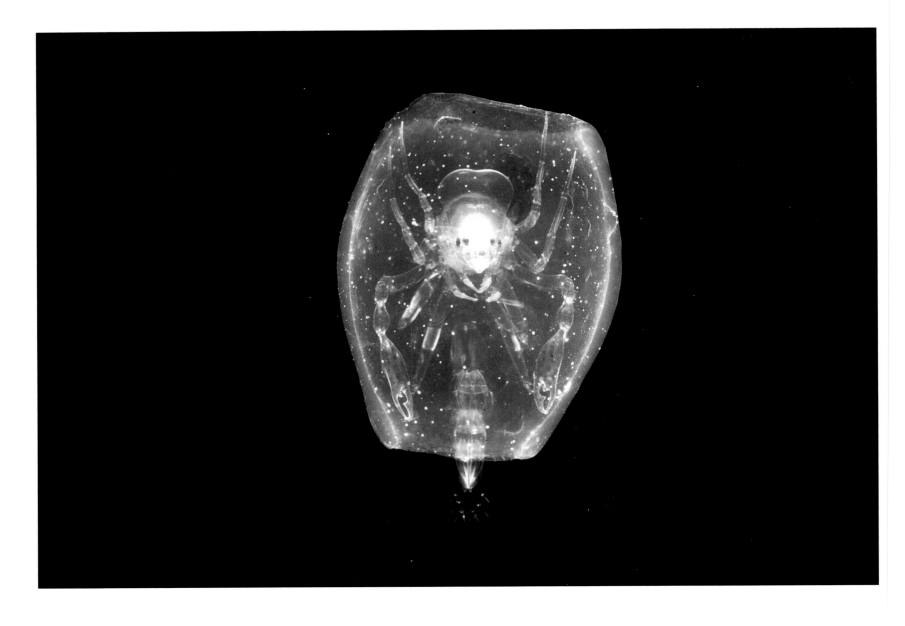

JUSTIN MARSHALL ~ CORAL SEA,
SOUTH PACIFIC OCEAN
*Sheltered in the abandoned hull of a translucent
salp, a* Phronima *amphipod drifts in deep water
near Osprey Reef in the Coral Sea.*

JUSTIN MARSHALL ~ CORAL SEA,
SOUTH PACIFIC OCEAN
Rarely seen alive, this Peraphilla *jellyfish prospers in
perpetual darkness 3,500 feet deep near Osprey Reef
in the Coral Sea.*

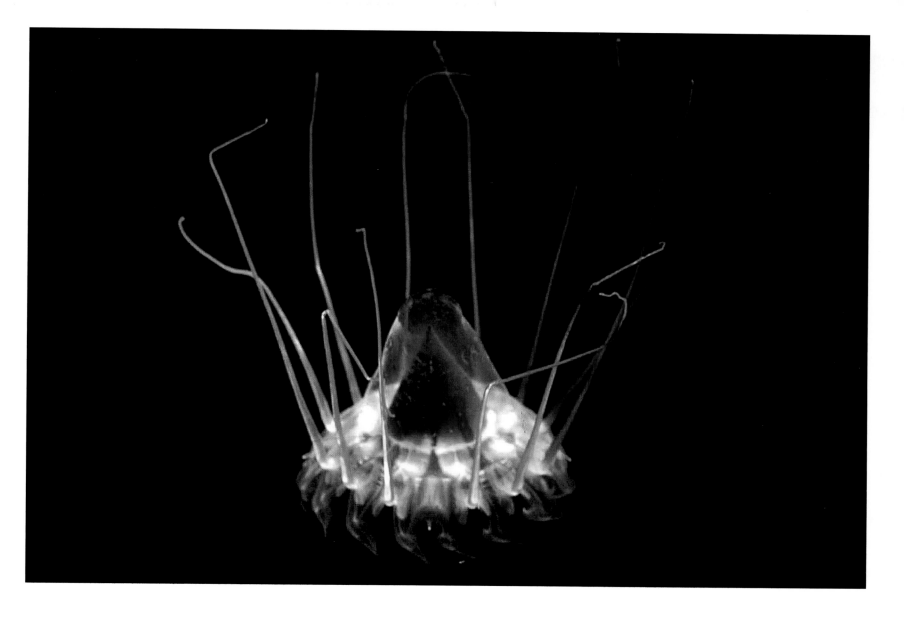

We tend to think of America's days of frontier
exploration as being behind us,
but that's because we tend not to think of
the other 71 percent of our blue planet.

—DAVID HELVARG, JOURNALIST & ENVIRONMENTAL ACTIVIST

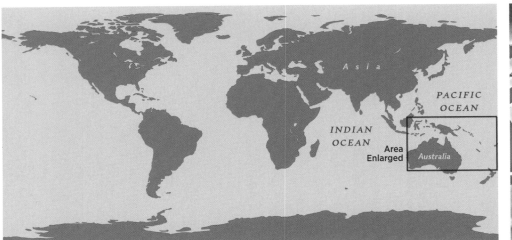

WEST CAROLINE BASIN

Halmahera

EQUATOR

0° Sulawesi

NEW GUI

North Banda Basin

Coral Triangle

Ceram

I N D O N E S I A

Kepulauan Aru

BANDA SEA

South Banda Basin

Weber

Aru Basin

LESSER SUNDA ISLANDS

Timor

ARAFURA SE

TIMOR TROUGH

TIMOR SEA

ARAFURA SHE

10°S

120°E

I N D I A N

NORTH AUSTRALIAN BASIN

O C E A N

SAHUL SHELF

Ca

ROWLEY SHELF

20°

A U S T R A L I A

TROPIC OF CAPRICORN

Miller Cylindrical Projection

SCALE 1:19,250,000

1 CENTIMETER = 192.5 KILOMETERS; 1 INCH = 303.8 STATUTE MILES

0   100   200   300

KILOMETERS

0   100   200   300

STATUTE MILES

0   100   200   300

NAUTICAL MILES

30°S

EUCLA SHELF

GREAT AUSTRALIAN BIGHT

120°E   130°

CORAL SEA & THE GREAT BARRIER REEF

The scene of fierce battle during World War II, the Coral Sea today is protected by Australia for its significant historic and natural values. Bounded to the west by the Great Barrier Reef (a World Heritage site), to the south by the Tasman Sea, to the east by Vanuatu and the Solomon Islands, and to the north by New Guinea, the Coral Sea hosts numerous seamounts, coral reefs, and deep ocean canyons that provide vital havens for ocean life—and hope for restoring depleted systems. About 386,100 square miles (a million sq km) was declared a commonwealth marine reserve by Australia in 2013, part of a comprehensive management plan for the nation's entire exclusive economic zone.

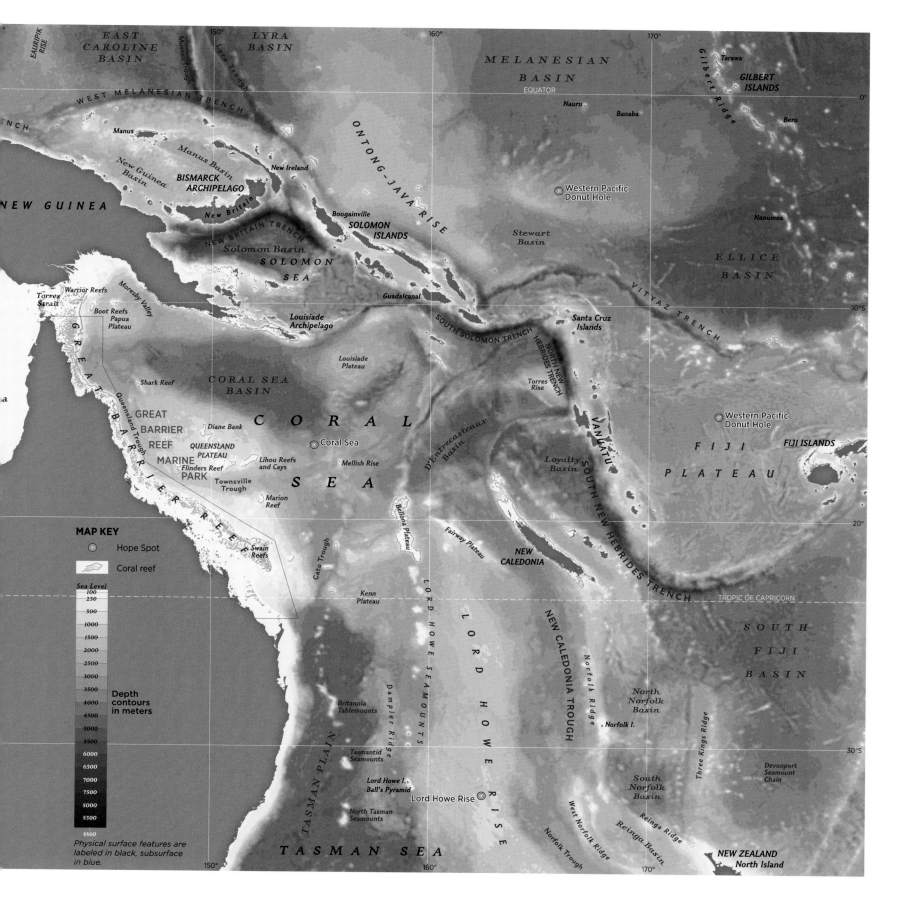

EAST CAROLINE BASIN

EAURIPIK RISE

LYRA BASIN

Mussau Trough

Lyra Trough

WEST MELANESIAN TRENCH

MELANESIAN BASIN

EQUATOR

Nauru

Banaba

Tarawa

GILBERT ISLANDS

Gilbert Ridge

Beru

Manus

New Guinea Basin

Manus Basin

BISMARCK ARCHIPELAGO

New Ireland

New Britain

ONTONG-JAVA RISE

Western Pacific Donut Hole

Nanumea

NEW GUINEA

NEW BRITAIN TRENCH

Solomon Basin

Bougainville

SOLOMON ISLANDS

SOLOMON SEA

Stewart Basin

ELLICE BASIN

Moresby Valley

Warrior Reefs

Torres Strait

Boot Reefs
Papua Plateau

Louisiade Archipelago

Guadalcanal

SOUTH SOLOMON TRENCH

Santa Cruz Islands

VITYAZ TRENCH

10°S

GREAT BARRIER REEF

Queensland Trough

Shark Reef

Louisiade Plateau

CORAL SEA BASIN

NORTH NEW HEBRIDES TRENCH

Torres Rise

GREAT BARRIER REEF

Diane Bank

C O R A L

Coral Sea

D'Entrecasteaux Basin

VANUATU

Western Pacific Donut Hole

GREAT
BARRIER
REEF
MARINE
PARK

QUEENSLAND PLATEAU

Flinders Reef

Lihou Reefs and Cays

Mellish Rise

S E A

Loyalty Basin

FIJI PLATEAU

FIJI ISLANDS

Townsville Trough

Marion Reef

Bellona Plateau

SOUTH NEW HEBRIDES TRENCH

MAP KEY

Hope Spot

Coral reef

Cato Trough

Fairway Plateau

NEW CALEDONIA

Swain Reefs

Sea Level
100
250
500
1000
1500
2000
2500
3000
3500
4000
4500
5000
5500
6000
6500
7000
7500
8000
8500

Depth contours in meters

8860

Kenn Plateau

LORD HOWE SEAMOUNTS

L O R D   H O W E   R I S E

NEW CALEDONIA TROUGH

Norfolk Ridge

Norfolk I.

TROPIC OF CAPRICORN

SOUTH FIJI BASIN

North Norfolk Basin

TASMAN PLAIN

Britannia Tablemounts

Dampier Ridge

Tasmantid Seamounts

Lord Howe I.
Ball's Pyramid

Lord Howe Rise

North Tasman Seamounts

Three Kings Ridge

West Norfolk Ridge

South Norfolk Basin

Reinga Ridge

Reinga Basin

Devonport Seamount Chain

30°S

Norfolk Trough

Physical surface features are labeled in black, subsurface in blue.

TASMAN SEA

NEW ZEALAND
North Island

TO THE
ENDS
OF THE
EARTH

~

We have one chance—now—
to protect or forever disrupt
deep polar seas never before
touched in the history
of humankind.

—SYLVIA A. EARLE

EXPLORING THE POLAR OCEANS

THE ARCTIC OCEAN & SOUTHERN OCEAN

Pitched out of my narrow bunk, I scrambled to hold on as the nuclear-powered Russian icebreaker *Sovietski Soyuz* plowed through ice seven feet thick while en route to the North Pole in 1998. On deck, bundled in several layers topped by a puffy red parka, I winced at the sight of our broad, trailing swath of dark water dotted with upended ice and shredded communities of golden diatoms and small animals that had been grazing on the underside of their frozen ceiling before being flipped into eternity.

The next day, on a starkly white expanse of snow-covered ice, I stood on the top of the world with a small group of fellow explorers at the one place on Earth where every direction is south. Among them was Alfred McLaren, retired captain of the U.S. Navy's nuclear submarine *Queenfish*, who had led the way through uncharted waters to achieve the first submerged survey of the Arctic Ocean's Canadian Northwest Passage in 1970, continuing into waters beyond, over the broad plains and jagged undersea peaks of the Mendeleev, Lomonosov, and Gakkel Ridges—rugged mountain chains that cut across deep basins within Earth's smallest, least known ocean.

From afar, Earth's polar regions glisten alabaster white, the frozen caps of a mostly blue planet, the only regions known where water occurs year-round as a liquid, solid, and vapor. To the north, the ice-covered Arctic Ocean is largely surrounded by land; to the south, the Southern Ocean races in a ragged circle around the Antarctic continent, an ice-covered landmass fringed with shelves of ever changing sea ice. Neither ocean has been accessible for most of human history, but development of new technologies and recent global warming have set in motion changes that have provoked unprecedented interest both in exploiting—and protecting—these unique regions.

For thousands of years, people have lived in the Arctic, but not until late in the 19th century was it confirmed what polar bears, walruses, whales, fish, seabirds, and legions of other aquatic animals of the region have always known: An ocean flows under the high Arctic's solid surface. Inspired by recovery of a shipwreck that had drifted across the Arctic embedded in pack ice, Norwegian explorer Fridtjof Nansen commissioned construction of a specially designed ship, *Fram,* and deliberately allowed her to become trapped into summer pack ice in the eastern Arctic in 1893. The character of the polar region as a deep, slowly circulating ice-covered ocean was confirmed when *Fram* and her crew emerged into the Atlantic near Greenland three years later.

Explorers who for centuries had looked for ways to transit the Arctic would be astonished at the thought of anyone doing so *under* the ice. Similarly, Adm. Robert Peary, who led an expedition in 1909 seeking to be the first to plant a flag at the North Pole, would surely be awed by the six Russian explorers who in 2007 dived in two small submarines through holes in the ice not far from where he stood and planted their titanium flag in the seafloor nearly three miles below.

At the other end of the planet, the Southern Ocean sweeps around Earth's highest, coldest, windiest, driest landmass, a continent twice the size of Australia that was not known to exist until

ARJUN GUPTA ~ NORWAY, ARCTIC OCEAN
Walruses bask on a beach behind Sylvia Earle at Svalbard, Norway, during a 2008 National Geographic expedition to the Arctic.

1820, when it was sighted by Russian explorers. Strong currents move water unimpeded around Antarctica, forming an effective liquid barrier separating the warmer waters of the Atlantic, Pacific, and Indian Oceans that converge at the edge of the much colder circumpolar current. More than 60 percent of the world's fresh water covers the Antarctic continent as a glittering slab of ice about two miles thick over more than five million square miles.

Largely ignored except by sealers, whalers, bird hunters, and fishermen until the middle of the 20th century, polar seas are presently visited annually by hundreds of hardy explorer-

the Southern Ocean is subjected to commercial extraction of marine wildlife, particularly krill and fish, under regulations developed by the Convention on the Conservation of Antarctic Marine Living Resources (CCAMLR), part of the Antarctic Treaty System that came into effect in 1982. Japan justifies killing hundreds of minke whales annually in the Southern Ocean under a research provision of the International Whaling Commission, although the area was established as the Southern Ocean Whale Sanctuary in 1994. CCAMLR condones the commercial taking of thousands of tons of krill, toothfish (marketed as Chilean sea bass), and other

> It makes sense, doesn't it, to carefully explore unknown regions of the sea before committing to irreversible exploitation? Perhaps we will discover that the most important value of vast parts of the ocean is already being extracted: our very existence.
>
> —SYLVIA A. EARLE

scientists and thousands of tourists aboard private and commercial cruise ships. Both regions are the topic of growing international controversy as new technologies overcome old obstacles to human access and warming melts away former barriers to extracting minerals, oil, gas, and marine life.

Antarctica, defined as the land and ice shelves south of 60 degrees south latitude, is the only continent without a native human population and has been remarkably protected from military and commercial use by the Antarctic Treaty System, a collection of international agreements initiated in 1959, now signed by 47 countries. The objective of the system is to ensure that Antarctica will serve the interests of all humankind for exclusively peaceful purposes.

Since no one nation "owns" Antarctica, the surrounding sea is part of the global commons, the internationally recognized "High Seas." On the continent, taking of wildlife is prohibited, but

marine life from the region. But there are questions about the impact large-scale extraction is having on previously isolated species and ecosystems that have no natural defenses against the technologically advanced methods used to find, capture, and market them.

The heart of the Arctic Ocean is also currently recognized as a global commons, beyond the exclusive economic zones of the five bordering nations. Sealed from human influence until recent decades, the icy barriers that have heretofore protected ancient, pristine systems are rapidly retreating. Perhaps the most startling discovery of Captain McLaren's Arctic undersea research aboard three nuclear subs in the 1960s and 1970s concerned not exploration of the terrain below, but the ice above. Compared to the first measurements of ice thickness made in 1958 from the nuclear submarine U.S.S. *Nautilus,* the thickness of the pack ice overall had thinned by 28 inches in 12 years. It is a trend that continues. Not only is Arctic sea ice becoming thinner, but also evidence gathered from ships and satellites shows dramatic shrinking in overall coverage, a consequence of global warming brought about by discharge into the atmosphere of carbon dioxide and methane as fossil fuels are burned to power increasing demands for energy.

By the middle of the 21st century, open-water passage across the Arctic during summer months is now predicted. This is good news for shipping, especially for traffic between Asia and Europe that now involves long and costly transits. It is also favorable for commercial extraction of oil, gas, minerals, and ocean wildlife. But for polar bears, walruses, life in the sea, the people of the Arctic— and of the world—rapidly melting ice means serious changes that will forever alter their lives—and livelihoods.

Measuring the wealth of polar seas in terms of what can be extracted is in keeping with traditional ways of valuing nature: tons of fish, barrels of oil, the furry hides of seals, foxes, wolves, and bears. However, this way of accounting does not recognize the value of the fundamental importance of natural systems to what humans tend to treasure the most: our existence. Earth's polar regions, sometimes characterized as the planet's air-conditioning system, have magnified significance in terms of driving ocean currents, shaping global climate and weather, governing temperature regimes within a range favorable to humankind. It is in our best interest to do everything in our power to maintain the integrity of these vital areas by reducing emissions of planet-warming gases and protecting their fabric of living systems from destructive actions.

Proposals for a network of protected areas in the Southern Ocean are being considered, with a special focus on the Ross Sea, a vital research area still largely intact despite recent large-scale extraction of krill and toothfish. A reevaluation of current policies for fishing in the High Seas region surrounding Antarctica could highlight the importance of keeping the natural systems as intact as possible in light of rapid changes that affect

YVA MOMATIUK & JOHN EASTCOTT ~ SOUTH GEORGIA ISLAND, SOUTHERN OCEAN
A young male Antarctica fur seal, choked by an abandoned fishing net on South Georgia Island in the Southern Ocean, is one of hundreds of thousands of marine mammals and birds that die every year from entanglement in fishing gear.

them—and connect back to humankind. For the High Seas portion of the Arctic Ocean, some measures are being taken by bordering countries to restrain commercial fishing, and support is growing for protective policies throughout the region.

Deliberate actions from now on will determine the fate of polar seas, one way or another. Elsewhere, choices have already been made, but the future of Earth's frozen assets will be determined in the next few years. In 2009, soon after I had attended scientific meetings to discuss the impacts of fishing in the Ross Sea, I attended the 80th birthday celebration in New York of revered Harvard biologist Edward O. Wilson. The audience was hushed when he remarked, "We are letting nature slip through our fingers." I shivered at the thought. Moreover, I wondered, "What if nature lets us slip through hers?"

Seeing is believing . . .
I was a climate change
skeptic until I saw the
evidence in the ice . . .
Climate change is real and
the time to act is now.

—JAMES BALOG, AUTHOR & EXPLORER

RALPH LEE HOPKINS ~ NORWAY, ARCTIC OCEAN
A polar bear and her cubs walk along the sparse
pack ice near the coast of Spitsbergen, an island
in the Svalbard Archipelago.

RALPH LEE HOPKINS ~ NORWAY, ARCTIC OCEAN
(*PRECEDING PAGES*)
Water in its three forms—liquid sea, solid ice,
and vaporous clouds—magically converges in
the Arctic Ocean near Svalbard, Norway.

While our polar areas are thawing, the consequent sea level rise is threatening the very existence of other regions far away.

—LISA-ANN GERSHWIN,
*STUNG! ON JELLYFISH BLOOMS
AND THE FUTURE OF THE OCEAN*

PAUL NICKLEN ~ NUNAVUT, CANADA,
ARCTIC OCEAN
*Drifting pack ice enables a walrus to remain
near a favored foraging area—a clam bed in
Foxe Basin, north of Hudson Bay.*

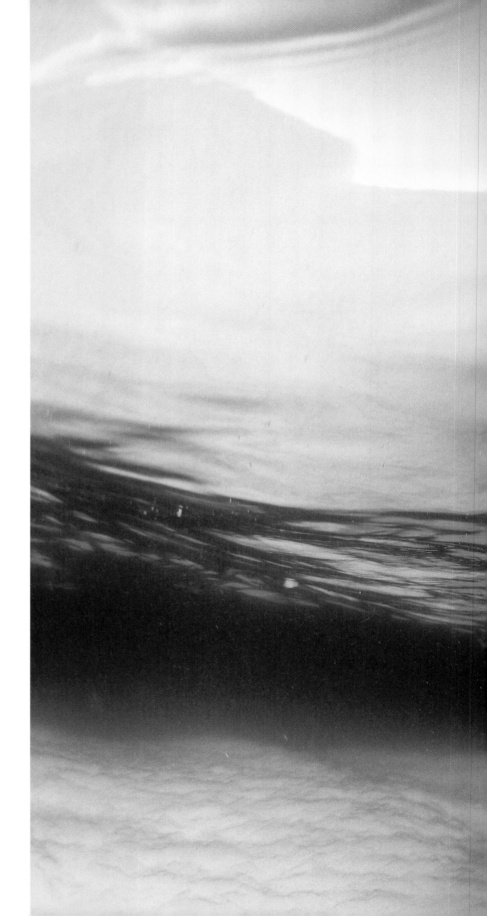

To waste, to destroy, our natural resources, to skin and exhaust the land instead of using it so as to increase its usefulness, will result in undermining in the days of our children the very prosperity which we ought by right to hand down to them amplified and developed.

—PRESIDENT THEODORE ROOSEVELT

PAUL NICKLEN ~ NORWAY, ARCTIC OCEAN
Beluga whale skulls and bones line the coast of Svalbard Archipelago—midway between Norway and the North Pole. Though greatly reduced from former abundance, these ivory-white "canaries of the sea" are still hunted in Canadian waters.

FLIP NICKLIN ~ NUNAVUT, CANADA, ARCTIC OCEAN
Clione, a pteropod or "sea butterfly," is a swimming cousin of snails and slugs, shown here near Baffin Island. Legions of these small animals dine on plankton and, in turn, are a critical food source for larger animals.

NORBERT WU ~ ARCTIC OCEAN
A pink helmet jellyfish is among the myriad planktonic creatures thriving in the deep, dark waters of the Arctic Ocean.

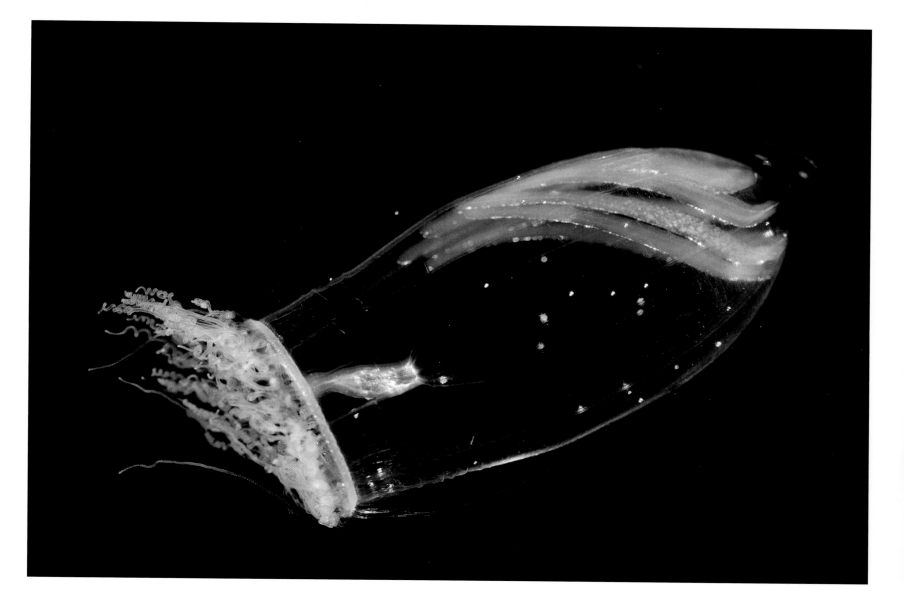

The only other place comparable
to these marvelous nether regions,
must surely be naked space itself . . .

—WILLIAM BEEBE, *NATIONAL GEOGRAPHIC,* 1934

The ocean was alive with little peaks . . . Every swell was born in a different place, made from a specific recipe of wind, time, and water . . . each wave was unique as a fingerprint. It had its own provenance and its own destiny, clashing against its neighbors or merging with them . . .

—SUSAN CASEY, *THE WAVE: IN PURSUIT OF THE ROGUES, FREAKS, AND GIANTS OF THE OCEAN*

TODD GIPSTEIN ~ ANTARCTICA, SOUTHERN OCEAN
Most of Earth's fresh water covers the Antarctic continent as a deep blanket of ice and in massive icebergs such as this one adrift near the Antarctic Peninsula in the Southern Ocean.

If some alien called me up . . .
"Hello, this is Alpha,
and we want to know what kind of life you have,"
I'd say, waterbased . . . Earth organisms figure out
how to make do without almost anything else.
The single nonnegotiable thing
life requires is water.

—CHRISTOPHER MCKAY, NASA SCIENTIST

FLIP NICKLIN ~ NUNAVUT, CANADA, ARCTIC OCEAN
A female beluga appears to glow in Cunningham
Inlet's surface waters near Somerset Island. Highly vocal,
belugas communicate with one another using musical
calls that resemble birdsong.

Sea of stretch'd
ground-swells!
Sea breathing broad
and convulsive breaths!
Sea of the brine of life! . . .
I am integral with you . . .

—WALT WHITMAN, "SONG OF MYSELF"

PAUL NICKLEN ~ ANTARCTICA, SOUTHERN OCEAN
*These emperor penguins are as at home in the
sea as they are on land. In search of fish, squid,
and small crustaceans, some dive as deep as
1,750 feet in the Southern Ocean.*

Yes! yes! give me this glorious ocean life, this salt-sea life, this briny, foamy life, when the sea neighs and snorts, and you breathe the very breath that the great whales respire!

—HERMAN MELVILLE, *REDBURN*

BRIAN SKERRY ~ NEW ZEALAND, SOUTHERN OCEAN
A southern right whale hovers inches above the sandy seafloor near New Zealand's Auckland Islands in the Southern Ocean. Individual identities are readily made owing to distinctive clusters of callosities that crown their faces.

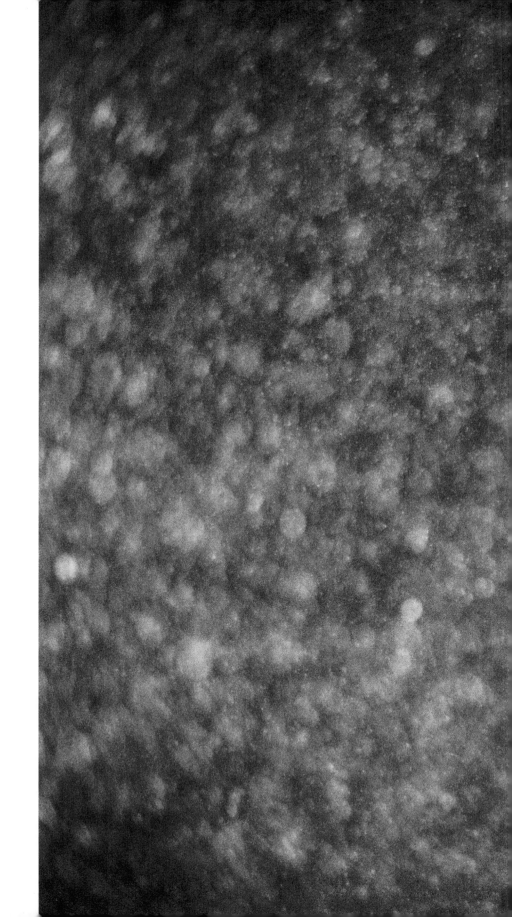

That great highway
extending from pole to pole,
which is forever closed
to human gaze, but may,
nevertheless, be penetrated
by human intelligence.

—GEORGE C. WALLICH,
THE NORTH ATLANTIC SEA BED

BILL CURTSINGER ~ ANTARCTICA, SOUTHERN OCEAN
*A fierce but curious predator, a leopard seal peers
through a veil of plankton. The last thing a penguin
might see is the face of a leopard seal, looming
through the misty depths of an icy sea. Ancient
rhythms of life link seals, penguins, fish, krill,
and plankton.*

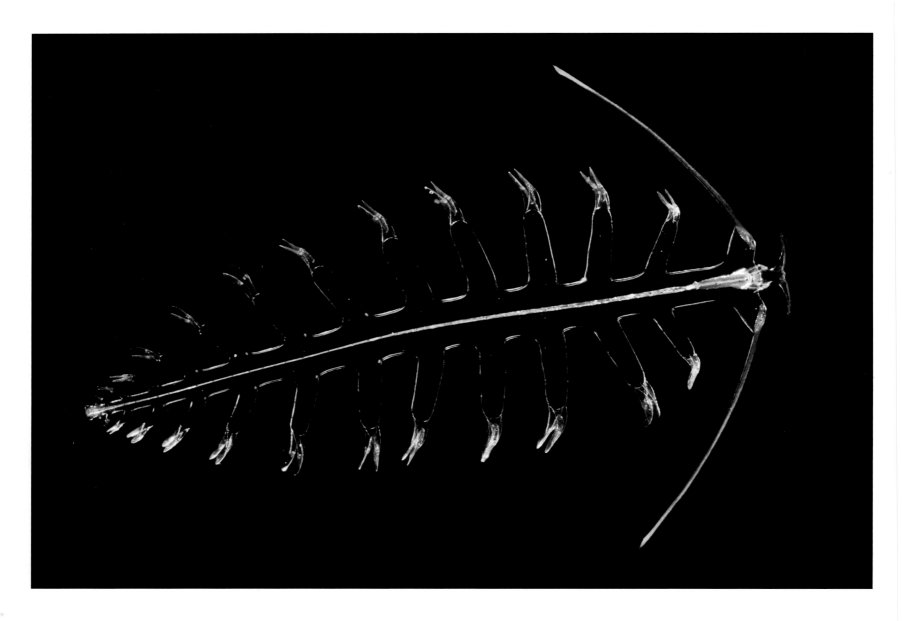

SOLVIN ZANKL ~ NORWAY, NORTH ATLANTIC OCEAN
*A bristleworm wends its way through the dark waters
of Trondheimsfjorden, an inlet of the Norwegian Sea.*

BILL CURTSINGER ~ ANTARCTICA, SOUTHERN OCEAN
*A thumbnail-size transparent jellyfish pulses deep within
McMurdo Sound.*

Whether it is to walk by her side,
swim in her cooling waters, or float on her surface,
the vast expanse of our ocean is transformative.
We stand in awe of her majesty.

—MARK SPAULDING, PRESIDENT, THE OCEAN FOUNDATION

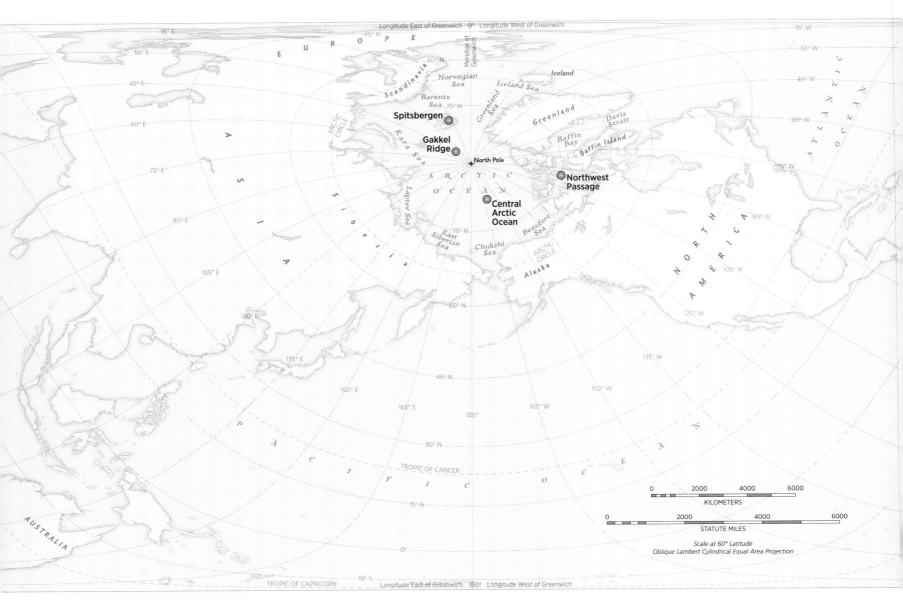

POLAR SEAS

Much of the Arctic is an ice-covered ocean largely surrounded by land while the Southern or Antarctic Ocean swirls unimpeded around the mostly ice-covered Antarctic continent. From afar, both polar areas appear as glistening whitecaps on a mostly blue planet, but underwater, the temperature remains above freezing and hosts an abundance and diversity of life that rival tropical coral reefs and forests. The Southern Ocean covers 8,097,843 square miles (20,973,318 sq km) with an average depth of 10,627 feet (3,239 m) and contains the largest current, the Antarctic Circumpolar, which transports more water than all of the world's rivers combined. The Arctic Ocean embraces 3,350,023 square miles (8,676, 520 sq km) with an average depth of 3,248 feet (990 m). Both regions, inaccessible to human activity until recent decades, are currently at the crossroads of decisions that will permanently affect their fate, and ours.

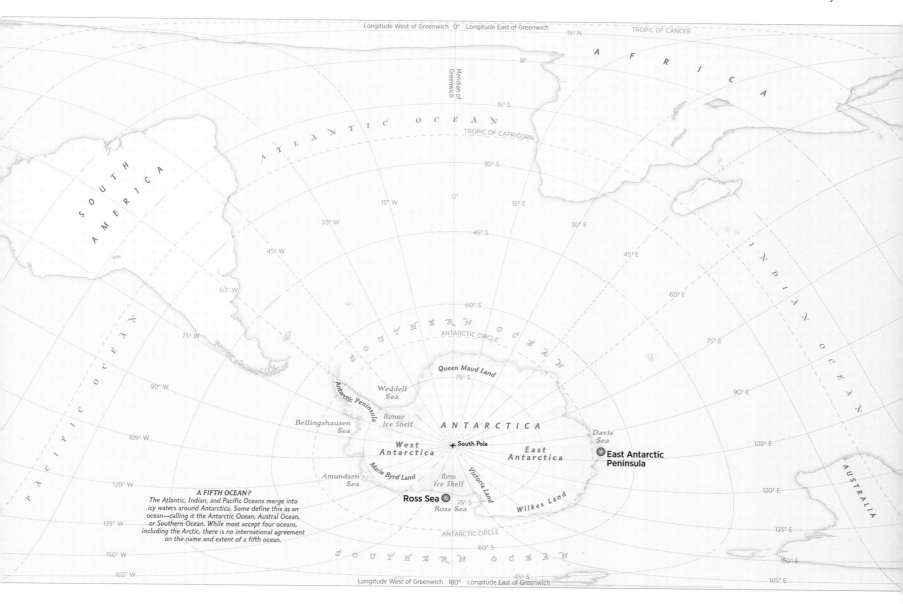

Longitude West of Greenwich 0° Longitude East of Greenwich

TROPIC OF CANCER

A F R I C A

Meridian of Greenwich

15° N

0°

15° S

ATLANTIC OCEAN

TROPIC OF CAPRICORN

30° S

SOUTH AMERICA

45° S

0°

15° W

15° E

30° W

30° E

INDIAN OCEAN

45° W

45° S

45° E

60° W

60° S

60° E

75° W

ANTARCTIC CIRCLE

75° E

SOUTHERN OCEAN

Queen Maud Land

75° S

90° W

PACIFIC OCEAN

Weddell Sea

Antarctic Peninsula

ANTARCTICA

Davis Sea

90° E

105° W

Bellingshausen Sea

Ronne Ice Shelf

East Antarctica

East Antarctic Peninsula

105° E

West Antarctica

+ South Pole

120° W

Amundsen Sea

Marie Byrd Land

Ross Ice Shelf

Victoria Land

120° E

AUSTRALIA

A FIFTH OCEAN?
The Atlantic, Indian, and Pacific Oceans merge into icy waters around Antarctica. Some define this as an ocean—calling it the Antarctic Ocean, Austral Ocean, or Southern Ocean. While most accept four oceans, including the Arctic, there is no international agreement on the name and extent of a fifth ocean.

Ross Sea ◎

Ross Sea

75° S

Wilkes Land

135° W

ANTARCTIC CIRCLE

135° E

150° W

SOUTHERN OCEAN

60° S

150° E

165° W

45° S

165° E

Longitude West of Greenwich 180° Longitude East of Greenwich

MAP KEY

◎ Hope Spot

SAVING PARADISE

~

With knowing comes caring, and with caring there is hope that we will find an enduring place for ourselves within the natural—mostly blue—systems that sustain us.

—SYLVIA A. EARLE

MAKING PEACE WITH THE OCEAN

GALÁPAGOS ISLANDS & THE GULF OF CALIFORNIA

Lulled to sleep by the humming of well-fed mosquitoes and the soft scrabbling of displeased fiddler crabs displaced by the dozen humans who had camped out on the crab's sand patch, I spent my second night in the Galápagos Islands in May 1966, among mangroves along Santa Cruz Island. The day before, the scientific party from cruise 16 of the research vessel *Anton Bruun,* of which I was a member, had improvised transport aboard a DC-4 aircraft to Ecuador's military base on Baltra Island, 600 miles offshore from Guayaquil, where the ship was undergoing vital repairs.

Bare bunks in barracks the first night were welcome as we began weeks of exploring what many know as "the Enchanted Islands"—a place where giant tortoises lumber on the slopes of active volcanoes; cormorants fly underwater with wings too small for aerial transport; lizards submerge to graze on algae; and tropical corals, mangroves, and pelicans share space with penguins, fur seals, and sea lions with Antarctic affinities. On day two, we hitched a ride with live chickens and other cargo on a small boat to Puerto Ayora, where the Charles Darwin Research Station had recently been established.

When the *Anton Bruun* arrived a few days later, we were provided with a compressor and sets of air tanks and regulators, enabling us to be among the first to explore dimensions of the Galápagos that English naturalist Charles Darwin missed during his historic 1835 visit—legions of hammerhead sharks, acres of garden eels dining on passing plankton, forests of kelp and red algae, and numerous kinds of subtidal invertebrates. In his field journal, Darwin wrote, "The main evil under which these islands suffer is the scarcity of water." Freshwater is limited, but the vast surrounding ocean of water makes the rich and diverse forms of life in the Galápagos possible, above and below the surface.

Sprawled across the Equator, the islands intercept the warm Panama current as it sweeps down from the north and two converging cold currents—the Peru, flowing from the east, and the subsurface Cromwell, flowing from the west. These create nutrient-rich water that is sharply cooler than the sun-warmed surface region. Often while diving in Galápagos waters, I encounter what seems to be the compression of thousands of miles of latitude within less than 200 vertical feet of ocean. From the waist up, I can share space with tropical angelfish, parrotfish, groupers, and great lumps of coral, while my flippers are chilled by cold water that enables forests of kelp and other cold-water organisms to prosper.

In 1966, about 1,000 people lived on 5 of the 18 major islands of the Galápagos Archipelago—mostly settlers who arrived in the 1920s and 1930s from Germany, Norway, and mainland Ecuador. Today, there are about 35,000 residents and nearly 200,000 annual visitors who come either to pay tribute to this internationally celebrated "living laboratory" or to experience the innocence of the wild animals—mockingbirds that fly to perch on your head, sea lions and iguanas that sleep peacefully when you stroll by, and boobies and albatrosses who dance to gain the attention of would-be mates. Underwater, sea lions frolic with snorkelers, and fish seem to accept

KIP EVANS ~ EASTERN PACIFIC OCEAN
Always armed with cameras to document observations in the ocean, Sylvia Earle explores the waters near Coiba, Panama.

divers as fellow sea creatures. So enamored have I been of such experiences that I have occasionally managed to extract each of my children from their academic formalities to attend schools of barracuda, rays, and angelfish; and in a burst of extravagance, I once managed to scoop all four grandsons, their moms, dads, and aunt for a week in the Galápagos aboard the National Geographic/Lindblad vessel *Endeavour.*

groupers, snappers, sea urchins, sea cucumbers, squid, octopus, lobsters, crabs, and whatever else could be eaten or marketed—was open to extraction, other than sea turtles and marine mammals, which were safeguarded by international policies.

But by 1986, depletion of marine species by an increasing number of residents and visitors and growing international and mainland Ecuador markets inspired the Ecuadorian government

> This is the sweet spot in time. As never before, and maybe as never again, there is a chance to protect the natural systems that keep us alive.
>
> —SYLVIA A. EARLE

The value of the extraordinary nature of the area was underscored in 1959, when Ecuador designated 97.5 percent of the Galápagos as a national park, excluding only areas already settled. Since the visit by Darwin 124 years before, actions by settlers, whalers, fishermen, and even some aggressive wildlife-collecting scientists had seriously disrupted the natural systems by burning unique landscapes and killing great numbers of whales, fish, tortoises, turtles, birds, iguanas, seals, and sea lions. Goats, pigs, cows, sheep, rats, cats, dogs, and numerous plant and insect species were also introduced. Protection as a park gave the natural systems a chance to recover and become more resilient to pressures imposed by periodic stresses of storms and El Niño and La Niña temperature shifts, which were brought about by changing currents.

At first, nothing was done to protect life in the sea around the Galápagos Islands. The reasons: a commonly held perception that the ocean takes care of itself and sea creatures are "seafood" to be freely captured by anyone. While all native birds, mammals, reptiles, plants, and invertebrates (except mosquitoes) were strictly protected, taking ocean wildlife—sharks,

to create a marine reserve—53,282 square miles around the islands. After years of conflict over fishing issues, guidelines for activities within the marine reserve were established in 1998. While the policies put in place were intended to protect marine life and sustain local interests, illegal fishing continued and demands by many—especially those interested in distant markets for sharks, lobsters, tunas, groupers, and sea cucumbers—trumped efforts to increase protection.

As a witness to life in the Galápagos that spans nearly half a century during dozens of expeditions, I am heartened to know that on the land, about 95 percent of the species present centuries ago are still there, although not in the same abundance and now sharing space with thousands of non-native species, including an inexorably increasing number of people. Underwater, several species of fish, algae, and invertebrates have apparently disappeared, likely because of disruptions caused by fishing coupled with several periods of prolonged warm temperature during El Niño years. Groupers, snappers, and other large fish are now uncommon; sea cucumbers and lobsters are scarce; and even the small fish critical as food for penguins and

other seabirds, seals, sea lions, dolphins, and whales are being swept up for use as bait. I once thought of the Galápagos Islands as the "sharkiest place on the planet," but decades of long-line fishing for mostly international markets have seriously depleted the sharks and other large ocean predators.

The next few years will determine whether or not the unique nature of the Galápagos Islands can be maintained or if the systems will gradually unravel as one species after another slips into oblivion. It is a choice, one that people, not penguins and iguanas, have the power to decide, and some recent actions indicate a trend toward protection.

In addition to Ecuador, three countries along the northeastern tropical Pacific coast have jurisdiction over offshore islands that have been given status as national parks and UNESCO World Heritage sites: Costa Rica's Cocos Island, located halfway between the Galápagos and the Costa Rican mainland; Colombia's Isla de Malpelo, a jagged exposed peak of a submerged seamount 300 miles offshore from Colombia's coast; and Panama's densely forested Coiba Island, once used as a penal colony but now a treasury of land and sea wildlife. Though very different on the surface, underwater these islands are safe havens for shared mobile populations of whales, turtles, sharks, tunas, and various other species.

Within the narrow Gulf of California between mainland Mexico and the Baja California peninsula, I found tangible evidence that protective measures can really work to protect healthy systems and restore depleted areas. As a visiting scientist at the California Academy of Sciences in the mid-1960s, I had dived during expeditions to various places along the Baja coast and reveled in encounters with yard-long groupers, great hammerhead sharks, and whirlpools of molten-silver sardines. I know what an intact ocean system looks like. And I know the desolation of places that have been stripped of essential elements.

Over the years, new houses, hotels, golf courses, and highways have transformed Baja wilderness to increasingly tamed terrain, and marine life has steadily decreased as great numbers of people have arrived, bringing their appetites with them. Commercial fishing, shrimping, and sportfishing have altered the ocean in parallel with more obvious changes to the land. But about 60 miles north of Los Cabos, Baja's epicenter of tourism, natural desert still dominates the landscape at Cabo Pulmo, a coastal community where fishing has long been a major occupation. But overfishing—too many people taking too much too fast—led to the collapse of fish populations, and therefore to fishing as a way of life.

In 1995, 27 square miles of ocean at Cabo Pulmo were declared a national park by the Mexican government, and revenues began to come in to local residents from snorkelers, divers, and others who cherish the broad beaches and quiet beauty of the coastal desert. Within ten years of protection, a striking recovery was under way. When I dived there in 2013, in the company of marine ecologists Jeremy Jackson, Octavio Aburto-Oropeza, and a team enlisted to document special places for the film *Mission Blue,* I glimpsed the Gulf of California that I had experienced 50 years ago. Shining through a wheeling tornado of jacks, dozens of parrotfish, and large dog-size groupers, I am certain that I saw *hope.* There is a chance to make peace with the ocean and secure an enduring future for tomorrow's children.

There is this extraordinary
connection between who
we are as human beings
and what happens in this
magnificent body of water.

—HILLARY RODHAM CLINTON,
"TURNING TO THE SEA: AMERICA'S OCEAN FUTURE"

**TIM LAMAN ~ GALÁPAGOS ISLANDS,
EASTERN PACIFIC ISLAND**
*Nimble acrobats underwater, Galápagos sea lions
are as placid as pillows while lounging on a sandy
beach in the Galápagos Islands.*

GEORGE STEINMETZ ~ BAJA PENINSULA, MEXICO
(PRECEDING PAGES)
*Winds sculpt sand into elegant patterns of sand
called barchan dunes at Ojo de Liebre.*

If there was ever a tide in humanity's relationship to the sea, it is now . . . Let us take this tide in its flood for the future of humanity and our wonderful planet!

—GRAEME KELLEHER, INTERNATIONAL UNION FOR CONSERVATION OF NATURE

CHRISTIAN VIZL ~ BAJA PENINSULA, MEXICO
An enormous school of jacks swims above a diver at Cabo Pulmo.

Who can say of a particular
sea that it is old?
Distilled by the sun, kneaded
by the moon, it is renewed
in a year, in a day, in an hour.

—THOMAS HARDY, *THE RETURN OF THE NATIVE*

DAVID DOUBILET ~ GALÁPAGOS ISLANDS,
EASTERN PACIFIC OCEAN
*A marine iguana perches on an underwater
rock, searching for green algae.*

BIRGITTE WILMS ~ GALÁPAGOS ISLANDS,
EASTERN PACIFIC OCEAN
*A large-banded blenny with a distinctive face hugs a rocky
perch 40 feet underwater in the Galápagos Islands. As with
other creatures, no two blennies are exactly alike.*

BIRGITTE WILMS ~ COSTA RICA,
EASTERN PACIFIC OCEAN
*Wearing a natural gloss, a rosy-lipped batfish
hunkers down on the seafloor near Cocos Island,
Costa Rica.*

She knew there was more to being a woman than
being kept in corsets. She wasn't content to float.
She was determined to swim.

—SWIMMER ESTHER WILLIAMS OF HER PREDECESSOR, ANNETTE KELLERMAN

The sea never changes, and
its works, for all the talk of
men, are wrapped in mystery.

—JOSEPH CONRAD, *FALK: A REMINISCENCE*

**FRED BAVENDAM ~ BRITISH COLUMBIA,
CANADA, NORTHEASTERN PACIFIC OCEAN**
*A Pacific giant octopus flings out its arms as
it settles onto the ocean bottom near Quadra
Island, British Columbia.*

Neither nature nor art has partitioned the sea into empires, kingdoms, republics, or states . . .

—PRESIDENT JOHN ADAMS

NORBERT WU ~ COSTA RICA,
EASTERN PACIFIC OCEAN
Hundreds of scalloped hammerhead sharks school around a seamount in Pacific waters near Cocos Island.

**DAVID DOUBILET ~ CALIFORNIA,
NORTH PACIFIC OCEAN**
*A hitchhiking crab clings to a purple-striped
jellyfish in Monterey Bay, California.*

**BILL CURTSINGER ~ BRITISH COLUMBIA, CANADA,
NORTHEASTERN PACIFIC OCEAN**
*Wearing a web of tiny hydroids, a sea snail prowls Bowie Seamount, a volcanic
undersea mountain in 10,000 feet of water that rises to within 80 feet of the surface.*

The wonders of the sea are as
marvelous as the glories
of the heavens.

—MATTHEW FONTAINE MAURY,
PHYSICAL GEOGRAPHY OF THE SEA

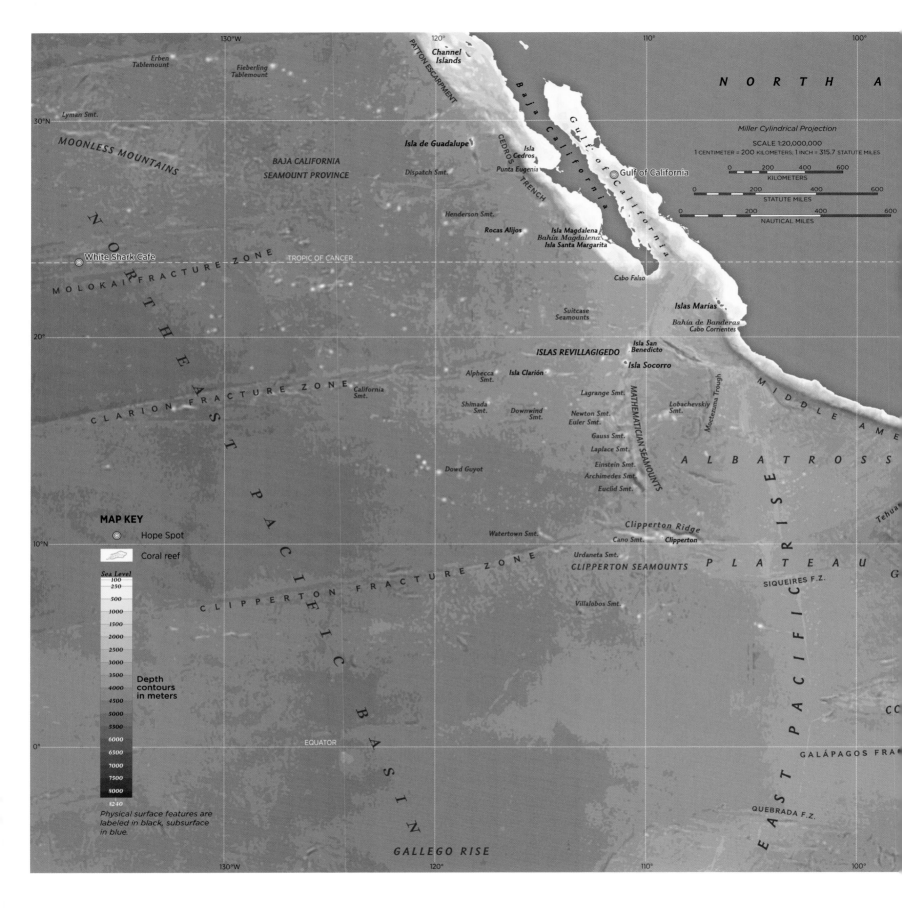

N O R T H   A

Channel
Islands

PATTON ESCARPMENT

Baja California

Gulf of California

30°N

Miller Cylindrical Projection

SCALE 1:20,000,000

1 CENTIMETER = 200 KILOMETERS; 1 INCH = 315.7 STATUTE MILES

Lyman Smt.

MOONLESS MOUNTAINS

Erben
Tablemount

Fieberling
Tablemount

Isla de Guadalupe

**BAJA CALIFORNIA
SEAMOUNT PROVINCE**

Dispatch Smt.

CEDROS TRENCH

*Isla
Cedros*
Punta Eugenia

◎ Gulf of California

0   200   400   600
KILOMETERS

0   200   400   600
STATUTE MILES

0   200   400   600
NAUTICAL MILES

N
O
R
T
H
E
A
S
T

Henderson Smt.

Rocas Alijos

Isla Magdalena
Bahía Magdalena
Isla Santa Margarita

White Shark Cafe ◎

TROPIC OF CANCER

MOLOKAI   FRACTURE ZONE

Cabo Falso

20°

F
R
A
C
T
U
R
E

Z
O
N
E

Suitcase
Seamounts

Islas Marías

Bahía de Banderas
Cabo Corrientes

CLARION   FRACTURE ZONE

California
Smt.

ISLAS REVILLAGIGEDO

*Isla San
Benedicto*

Alphecca
Smt.

Isla Clarión

Isla Socorro

Lagrange Smt.

Lobachevskiy
Smt.

MATHEMATICIAN SEAMOUNTS

Moctezuma Trough

MIDDLE   AME

Shimada
Smt.

Downwind
Smt.

Newton Smt.
Euler Smt.

A L B A T R O S S

Gauss Smt.

Laplace Smt.

Dowd Guyot

Einstein Smt.

Archimedes Smt.

Euclid Smt.

E
A
S
T

P
A
C
I
F
I
C

MAP KEY

◎   Hope Spot

Coral reef

Clipperton Ridge

Watertown Smt.

Cano Smt.

Clipperton

P L A T E A U

E
A
S
T

P
A
C
I
F
I
C

R
I
S
E

Tehua

10°N

Urdaneta Smt.

CLIPPERTON SEAMOUNTS

SIQUEIRES F.Z.

CLIPPERTON   FRACTURE   ZONE

Villalobos Smt.

Sea Level
100
250
500
1000
1500
2000
2500
3000
3500
4000
4500
5000
5500
6000
6500
7000
7500
8000
8240

**Depth
contours
in meters**

B
A
S
I
N

EQUATOR

GALÁPAGOS FRA

0°

*Physical surface features are
labeled in black, subsurface
in blue.*

CO

GALLEGO RISE

QUEBRADA F.Z.

130°W   120°   110°   100°

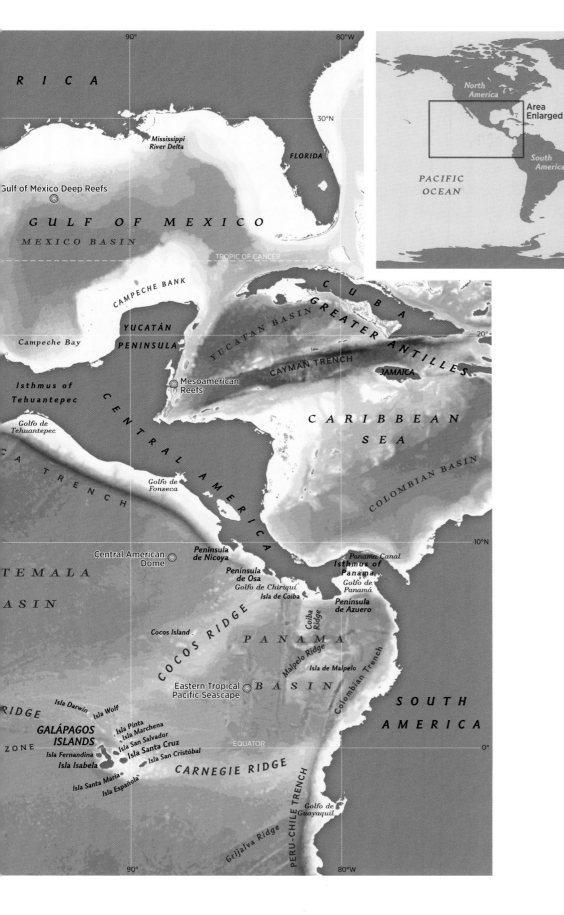

GALÁPAGOS ISLANDS & THE GULF OF CALIFORNIA

Waters along the southwestern United States, Mexico's Baja California, and coastal Central and South America south to Ecuador border eight countries with political boundaries and exclusive economic zones (EEZs) extending seaward 200 miles (322 km). The area around offshore islands greatly increases each nation's jurisdiction. For Costa Rica, the EEZ along the coasts and offshore Cocos Island magnifies the country tenfold. For Ecuador, with the EEZ of the Galápagos Islands included, oceanic jurisdiction is four and a half times larger than the land. Nations along this eastern tropical Pacific seascape are collaborating to manage activities with special reference to marine mammals, turtles, tunas, and other fish that range widely across transboundary corridors.

EPILOGUE

REASONS FOR HOPE

In response to the chance in 2009 to make a TED wish "big enough to change the world," a man previously unknown to me quietly provided the means to gather a ferment of scientists, explorers, artists, musicians, philanthropists, writers, and leaders in conservation, industry, and government. The goal: to explore ways to create the wished-for campaign, foster a network of marine protected areas, and build new submarines to explore Earth's blue heart—the ocean. In April 2010, a hundred individuals set sail for the first ever TED-at-sea conference. Our destination: the Galápagos Islands.

Aboard the National Geographic/Lindblad vessel *Endeavour,* TED-style talks and deep conversations alternated with walks along black, green, and golden sand beaches; scrambles over glassy-edged lava; and for some, first-time experiences seeing live fish underwater, face to face. At the outset, we were challenged to generate ideas and figure out how to implement tangible results—before leaving the ship. Some agreed to work together to identify and fund priority areas for protection; some provided help to enforce fishing laws in the Galápagos. The Sargasso Sea Alliance was initiated, and the concept was born for bringing together notable individuals as "ocean elders." A group formed around educating children inspired by the concept of "no child left dry," while others focused on protection for the High Seas and polar regions. A film team began documenting the concept of "hope spots." Carl Lundin and Dan Laffoley, ocean experts with the International Union for Conservation of Nature (IUCN), pledged their support for identifying critical ocean areas. New ideas continue to be spawned, each one sparking more.

A nonprofit foundation, the Sylvia Earle Alliance/Mission Blue was formed with initial collaboration from the National Geographic Society, now engaging nearly 100 partner organizations that are featured on the Mission Blue website, *mission-blue.org.*

Since 2010, the amount of ocean fully protected has more than doubled, but remains at less than one percent. Managed areas have doubled as well, now embracing close to 3 percent of the ocean—still a long way from the Mission Blue goal of 20 percent by 2020, but recent actions are promising. Pacific island nations such as Palau, Kiribati, and the Cook Islands are already changing the dynamic of exploitation to protection, and if Australia stays the course for protection of 30 percent of its exclusive economic zone, other nations are likely to follow. The Sargasso Sea Alliance, initiated during the 2010 Galápagos expedition, brought together representatives of more than a dozen nations in Bermuda early in 2014, and five signed the "Hamilton Declaration," pledging their support for conservation of two million square miles of the High Seas home of vast floating "forests" of *Sargassum.* At the 2013 International Marine Protected Areas Congress in Marseille, IUCN and Mission Blue launched an updated map of 50 Hope Spots that are supported by scientific analysis, and more will be added in 2014 and beyond. Organizations are pulling together to secure protection for critically important areas in the ocean. Twenty percent by 2020 is possible.

Wisdom, the Laysan albatross who first took flight more than six decades ago, has not stopped producing chicks just because the odds of their survival is slim. There is reason for hope, and hope is the reason success is possible.

ANN BELL ~ MIDWAY ATOLL
Wisdom, a Laysan albatross banded at Midway Island in the 1950s, admires her latest chick, "Hope," hatched nearby in 2014.

MISSION BLUE'S
HOPE SPOTS

~

Fish, whales, clams, krill,
and other ocean wildlife
have an accounting base of zero
when they are alive.
They are regarded as free
for the taking with nothing subtracted
from the ocean's balance sheet
when they are captured.
But there is no free lunch.
All of nature and all of us pay,
one way or another,
for what is extracted.

—SYLVIA A. EARLE

Hope Spots are areas in the ocean that have been identified for protection owing to their magnified importance in protecting, restoring, and maintaining the health of the ocean. Fifty-one Hope Spots are listed below, and more are being added as new areas and opportunities are identified. Some are large, some small; some are pristine, whereas others are greatly stressed. But with care, all provide hope for steady improvement in the current state of ocean health and resilience. According to analyses by the International Union for Conservation of Nature, about 12 percent of Earth's terrestrial area is designated for parks, reserves, and wildlife management zones, but for the sea, only about 2.5 percent has equivalent care. A fraction of one percent of the ocean is fully protected. For successful recovery of ocean species and systems, much more is needed. Mission Blue is committed to protecting the ocean, Earth's blue heart—and in so doing, provides hope for an enduring future for humankind.

1 CHAGOS ARCHIPELAGO, INDIAN OCEAN
The archipelago consists of 55 low-lying coral islands that span 212,356 square miles (550,000 sq km). Chagos's waters are home to Earth's largest coral atoll.

2 OUTER SEYCHELLES, INDIAN OCEAN
The Outer Seychelles—off Africa's coast—are a collection of five coralline island groups that include 72 low-lying sand cays and atolls.

3 CORAL TRIANGLE, WESTERN PACIFIC OCEAN
The Coral Triangle is a global center for biodiversity and is considered by many to be the most diverse marine ecosystem in the world.

4 MICRONESIAN ISLANDS, PACIFIC OCEAN
The Micronesian Islands—2,100 tropical islands scattered across the heart of the Pacific—offer some of Earth's most pristine and biodiverse underwater environments.

5 CORAL SEA, OFF THE NORTHEAST COAST OF AUSTRALIA
This marginal sea is named for its staggering number of corals. The area includes the Great Barrier Reef and is one of the most diverse marine habitats on Earth.

6 KERMADEC TRENCH, SOUTH PACIFIC OCEAN
The Kermadec Trench is a submarine trench that plunges more than 6.2 miles (10 km) beneath the ocean's surface—about five times deeper than the Grand Canyon.

7 GULF OF CALIFORNIA, ALONG THE NORTHWEST COAST OF MEXICO
The Gulf of California, aka the Sea of Cortés, is a large inlet that covers

2,485 miles (4,000 km) of coastline and reaches depths of more than 9,840 feet (3,000 m).

8 GULF OF MEXICO DEEP REEFS
These 200 shelf-edge reefs and banks, hot spots along the continental shelf, support an abundance of soft corals, subtropical and tropical invertebrates, and 90-plus species of fish.

9 PATAGONIAN SHELF, OFF THE COAST OF ARGENTINA
The shelf is part of the South American continental shelf. The southward flowing Brazil Current, warm and saline, mixes with the northward flowing Falklands Current, cool and less saline.

10 EASTERN TROPICAL PACIFIC SEASCAPE
The seascape spans the west coast of Central and South America. Some of Earth's most important natural habitats occur here, including the Galápagos Islands.

11 CHILEAN FJORDS AND ISLANDS, CHILE
Chile's marine territory covers more than 52,100 miles (83,850 km) of coastline and includes several islands. The fjords offer important habitats for whales, dolphins, and other marine mammals.

12 ROSS SEA, SOUTHERN OCEAN
The Ross Sea is often referred to as Earth's most pristine marine ecosystem. It remains relatively free from pollution, invasive species, mining, and overfishing.

13 MESOAMERICAN REEFS, CARIBBEAN SEA
The reefs—home to more than 350 species of mollusk and 500

species of fish—extend from Isla Contoy on the northern Yucatán Peninsula to the Bay Islands of Honduras.

14 GULF OF GUINEA, OFF THE WEST COAST OF AFRICA
The gulf is part of the eastern tropical Atlantic Ocean. Its marine flora and fauna are largely intact; the area is considered critical to preserving African biodiversity.

15 CHARLIE-GIBBS FRACTURE ZONE, NORTH ATLANTIC OCEAN
The fracture zone is a major transversal topographical feature located beyond the limits of national jurisdiction. It reaches depths of 14,764 feet (4,500 m).

16 SARGASSO SEA, NORTH ATLANTIC OCEAN
The Sargasso Sea is Earth's only sea without a land boundary. The ecosystem is bounded by currents circulating around the North Atlantic subtropical gyre.

17 GAKKEL RIDGE, ARCTIC OCEAN
The Gakkel Ridge ecosystem is thought to host diverse life found nowhere else on Earth—in particular, those areas surrounding the ridge's hydrothermal vents.

18 BAHAMIAN REEFS, OFF SOUTHEASTERN FLORIDA
This region consists of more than 3,000 low-lying islands. It hosts forests, wetlands, swamps, and the Andros Barrier Reef, the second largest barrier reef in the Western Hemisphere.

19 BERING SEA DEEP CANYONS, BERING SEA
Home to ocean albatross and kittiwakes, orcas, walrus, king crab, salmon, and cold-water corals, the Bering Sea Canyons support a near endless variety of life.

20 CENTRAL ARCTIC OCEAN
This area is the last unfished Arctic high seas enclave. While currently a nutrient-poor area, the disappearance of ice and other climate-related impacts makes it vulnerable to overfishing.

21 WHITE SHARK CAFÉ, NORTHEASTERN PACIFIC OCEAN
The White Shark Café is an area of seasonal aggregation for endangered great white sharks, which come from two coastal wintering areas—central California and Guadalupe Island, Mexico.

22 SALA Y GÓMEZ AND NAZCA RIDGES, SOUTHEASTERN PACIFIC OCEAN
The ridges are sequential chains of submarine mountains of volcanic origin. Currently, 226 species of invertebrates and 171 fish species of 64 genera are known to inhabit the explored seamounts there.

23 TASMAN SEA, OFF THE COAST OF TASMANIA
The vulnerable antipodean albatross from the Auckland Islands forages over the Tasman Sea. The area is also used by at least five other species of threatened seabirds.

24 WALTERS SHOAL, SOUTHERN INDIAN OCEAN
This large seamount has been targeted by deep-sea and lobster fishing. The shoal appears to be a unique feature in the ocean there, forming part of the Madagascar Ridge.

25 CORAL SEAMOUNT, SOUTHWEST INDIAN RIDGE
This seamount hosts rich coral communities and cold-water coral reefs. Named an ecologically and biologically significant area (EBSA), it is also a voluntary benthic protection area (BPA) from the fishing industry.

26 ATLANTIS BANK, SOUTHWEST INDIAN RIDGE
The tectonic seamount is home to significant populations of pelagic armorhead, large octocorals, and sponges. The flat-topped seamount is also a voluntary BPA.

27 AGULHAS FRONT, SOUTHWESTERN INDIAN OCEAN
A key feeding area for the critically endangered Amsterdam albatross, the front is also key for at least three other threatened albatross species that breed in the French southern territories.

28 CORE OF THE SOUTH PACIFIC GYRE, SOUTHERN PACIFIC OCEAN
This area, which encompasses the Galápagos Rise, is one of two potential high-use regions for post-nesting Eastern Pacific leatherback females within the South Pacific Gyre.

29 PACIFIC SUBTROPICAL CONVERGENCE ZONE, SOUTHERN PACIFIC OCEAN
This zone, which encompasses the Challenger Fracture Zone, is one of two potential high-use regions for post-nesting Eastern Pacific leatherback females within the South Pacific Gyre.

30 FRENCH OVERSEAS TERRITORIES, SOUTH PACIFIC OCEAN
This area consists of 20 islets and three main volcanic islands. Five types of toothed whales inhabit the waters. Twenty seabird species have known distributions here.

31 LORD HOWE RISE, OFF THE EAST COAST OF AUSTRALIA, SOUTH PACIFIC
From the depths of Lord Howe Rise, a mountain rises to form Lord Howe Island, whose surrounding waters harbor 100-plus species found nowhere else on Earth.

32 ATOLLS OF THE MALDIVES, INDIA
A species-rich marine area, the islands contain atolls with natural freshwater lakes and wetlands that are refuges for diverse populations of birds, fish, and fauna.

33 LAKSHADWEEP ISLANDS, INDIA
The reefs here are an important biogeographic link between the

subcontinent and East Africa. The waters are relatively high in marine mammal diversity.

34 ANDAMAN ISLANDS, INDIA
The Andaman Islands—about 325 volcanic islands—are characterized by high-diversity fringing reefs with upward of 200 species of coral, extensive mangrove forests, and sea grass meadows.

35 SUBANTARCTIC ISLANDS AND SURROUNDING SEAS, SOUTHERN OCEAN
This area is among Earth's least human-modified environments. Threatened species include the New Zealand sea lion, several endemic species of land birds and seabirds, and two penguin species.

36 SAYA DE MALHA BANKS, WESTERN INDIAN OCEAN
The banks are part of the Mascarene Plateau. Considered unique due to their geological origin, they also represent the largest shallow open ocean-water biotope in the region.

37 SOUTHEAST SHOAL OF THE GRAND BANKS, NORTH ATLANTIC OCEAN
The Grand Banks is a shallow, submerged extension of Newfoundland that juts into the North Atlantic. It has a long history of fishing.

38 EMPEROR SEAMOUNT CHAIN, NORTH PACIFIC OCEAN
The volcanic chain arose from molten rock in the ocean between the Hawaiian and Aleutian Islands. Albatrosses, whales, and tunas visit the nutrient-rich waters to feed.

39 EAST ANTARCTIC PENINSULA, ANTARCTICA
Penguins, seals, and krill rely on this expanse of frigid habitat—as well as snow petrels, Antarctic terns, and albatrosses. The area also supports minke, humpback, blue, and fin whales.

40 GRAND RECIF DE TOLIARA, MADAGASCAR
Five hundred reef fish species have been recorded here, as well as a high diversity of other marine life—along with significant stands of mangroves and extensive sea grass beds.

41 CENTRAL AMERICAN DOME, NORTHEASTERN PACIFIC OCEAN
The dome, in the northeastern tropical Pacific, supports marine predators such as tuna, dolphins, and cetaceans. The endangered leatherback turtle migrates through here.

42 WESTERN PACIFIC DONUT HOLES
The area includes four High Seas enclaves: West Oceania Marine Reserve, Greater Oceania Marine Reserve, Moana Marine Reserve, and Western Pacific Marine Reserve.

43 SCOTT ISLANDS, CANADA
More than 400,000 pairs of Cassin's auklets nest on Triangle Island in the Scott Islands. The island also hosts 40,000 pairs of rhinoceros auklets and 26,000 pairs of tufted puffins.

44 NORTHWEST PASSAGE
The area is home to the largest narwhal congregation on Earth, and one-seventh of the world's beluga population. Several million migratory birds congregate here.

45 ABROLHOS BANK, SOUTH ATLANTIC OCEAN
The Abrolhos region is a mosaic of marine and coastal ecosystems that encompass the largest reef area and highest marine biodiversity in the southern Atlantic.

46 CHILOÉ NATIONAL PARK EXPANSION, CHILE
An important area for cetaceans, seabirds, sea lions, and marine otters, the park is largely inaccessible with no significant human impact.

47 ASCENSION ISLAND, SOUTH ATLANTIC OCEAN
Much of Ascension is a wasteland of lava flows and cinder cones. Even so, the island harbors globally important marine biodiversity, including bottlenose dolphins and humpback whales.

48 KOSTERFJORDEN/YTRE HVALER, NORWAY/SWEDEN
The area hosts a very high diversity of marine species and contains unique habitats and species found nowhere else in Sweden or this part of Norway.

49 MERGUI ARCHIPELAGO, SOUTHERN MYANMAR, ANDAMAN SEA
The archipelago consists of 800-plus islands. Its isolation from humankind has given the islands and surrounding waters a great diversity of flora and fauna.

50 QUIRIMBAS ISLANDS, INDIAN OCEAN OFF NORTHEASTERN MOZAMBIQUE
The archipelago consists of about 27 islands. Coral reef species typical of the region are visible, as well as manta rays, whale sharks, green and hawksbill turtles, and humpback whales.

51 SPITSBERGEN, NORWAY
Spitsbergen is the largest island of the Svalbard Archipelago. It is a breeding ground for many seabirds; it also supports polar bears, reindeer, and marine mammals.

ACKNOWLEDGMENTS

Deepest admiration and thanks to all who have made this tribute to the ocean possible—especially the editors, designers, and staff of National Geographic Books, with superlative creative skills and admirable teamwork, led by project manager Barbara Payne, art director Sanáa Akkach, photograph editor Matthew Propert, and map master Carl Mehler. For their sustained, positive vision concerning the merits of this book, I am lovingly grateful to Barbara Brownell Grogan and Bill O'Donnell. For the Hope Spots maps and underlying information, I especially thank Dr. Dan Laffoley, marine vice chair for IUCN's World Commission on Protected Areas, and Dr. Ben Halpern and his team at Duke University's Nicholas School of the Environment. I also must give a deep bow of respect to the contributing photographers, who have used their talents to capture the essence of the ocean's blue heart, conveyed in ways that transcend words.

On behalf of all who care about the ocean, I want to offer a special salute to La Mer, for its admirable corporate ethic of caring for the Earth, and for the support it provided to make this volume possible. I am personally grateful to the National Geographic Society, especially Terry Garcia and NGS Mission programs, for backing me as explorer-in-residence; to Chris Anderson and TED for granting me a "wish big enough to change the world"; to the Mission Blue staff and partners, and the team of Mission Blue filmmakers, led by Robert Nixon and Fisher Stevens, who have worked closely with me during the production of this volume.

I am grateful for the patience of family and friends who have suffered benign neglect for the time missed with them as this book was being crafted, especially my children—Gale Mead, Elizabeth Taylor, and Richie Mead; their spouses Ian Griffith and Tamara Miller; and grandsons, Russell Mead, Kevin Mead, Taylor Griffith, and Morgan Griffith. For them and for all of tomorrow's children, this book is lovingly dedicated.

ABOUT THE AUTHOR

National Geographic Explorer-in-Residence Sylvia A. Earle, named *Time* magazine's first Hero of the Planet and a Library of Congress Living Legend, is leader of the NGS Sustainable Seas Expeditions, founder of Mission Blue/Sylvia Earle Alliance, founder of Deep Ocean Exploration and Research, and former chief scientist of NOAA. Author of more than 200 publications, Earle has led 100-plus expeditions and lectured in more than 80 countries. She is a graduate of Florida State University and holds a master's degree and a Ph.D. from Duke University. Her research concerns the ecology and conservation of marine ecosystems and development of technology for access to the deep sea.

MAP SOURCES

Amante, C., and B. W. Eakins, *ETOPO1 1 Arc-Minute Global Relief Model: Procedures, Data Sources and Analysis.* NOAA Technical Memorandum NESDIS NGDC-24, 19 pp, March 2009. ngdc.noaa.gov/mgg/global/global.html

CLEANTOPO2/Tom Patterson: shadedrelief.com/cleantopo2

Duke University Marine Geospatial Ecology Lab: mgel.env.duke.edu

IUCN and UNEP-WCMC (2013), The World Database on Protected Areas (WDPA), December 2013 release. Cambridge, UK: UNEP-WCMC. wdpa.org

Marine Conservation Institute: marine-conservation.org

Mission Blue/Sylvia Earle Alliance: mission-blue.org

Sargasso Sea Alliance: sargassoalliance.org

Global Distribution of Coral Reefs dataset:

• IMaRS-USF (Institute for Marine Remote Sensing-University of South Florida) (2005). Millennium Coral Reef Mapping Project. Unvalidated maps. These maps are unendorsed by IRD, but were further interpreted by UNEP World Conservation Monitoring Centre. Cambridge (UK): UNEP World Conservation Monitoring Centre.

• IMaRS-USF, IRD (Institut de Recherche pour le Developpement) (2005). Millennium Coral Reef Mapping Project. Validated maps. Cambridge (UK): UNEP World Conservation Monitoring Centre

• Spalding MD, Ravilious C, Green EP (2001). World Atlas of Coral Reefs. Berkeley (California, USA): The University of California Press. 436 pp.

• UNEP-WCMC, WorldFish Centre, WRI, TNC (2010). Global distribution of warmwater coral reefs, compiled from multiple sources (listed in "Coral_Source.mdb"), and including IMaRS-USF and IRD (2005), IMaRS-USF (2005) and Spalding et al. (2001). Cambridge (UK): UNEP World Conservation Monitoring Centre. URL:www.data.unep-wcmc.org/datasets/13

VLIZ (2014). Maritime Boundaries Geodatabase, version 8. Available online at marineregions.org. Consulted April 2014.

Individual thanks to: Jesse Cleary, Dan Laffoley, Amy Milam, Kate Killerlain Morrison

ILLUSTRATIONS CREDITS

Front Cover, Francis Perez; Back Cover, Brian J. Skerry/National Geographic Creative; Author Photo, Bryce Groark; 2-3, Flip Nicklin/Minden Pictures/National Geographic Creative; 4-5, Brian Skerry/National Geographic Creative; 6-7, Brian Skerry/National Geographic Creative; 8-9, Maria Stenzel/National Geographic Creative; 10-11, Fred Bavendam/Minden Pictures/National Geographic Creative; 12-13, Flip Nicklin/National Geographic Creative; 14, Brian Skerry/National Geographic Creative; 20, Susan Middleton; 23, David Liittschwager/National Geographic Creative; 24-25, Frans Lanting/National Geographic Creative; 26-27, Frans Lanting/National Geographic Creative; 28-29, Flip Nicklin/Minden Pictures/National Geographic Creative; 30, Chris Newbert/Minden Pictures/National Geographic Creative; 31, David Doubilet/National Geographic Creative; 32-33, Brian Skerry/National Geographic Creative; 34, Chris Newbert/Minden Pictures/National Geographic Creative; 35, Chris Newbert/Minden Pictures/National Geographic Creative; 36-37, Mattias Klum/National Geographic Creative; 38, Michael Aw; 39, David Wrobel/Visuals Unlimited/Corbis; 44, Kip Evans; 47, Joel Sartore/National Geographic Creative; 48-49, Brian Skerry/National Geographic Creative; 50-51, David Doubilet/National Geographic Creative; 52-53, Brian Skerry/National Geographic Creative; 54-55, Brian Skerry/National Geographic Creative; 56, Paul Sutherland/National Geographic Creative; 57, David Doubilet/National Geographic Creative; 58-59, Bryce Groark; 61, Sandra Critelli; 62-63, Paul Nicklen/National Geographic Creative; 64, David Fleetham/naturepl.com; 65, Brian Skerry/National Geographic Creative; 66-67, David Doubilet/National Geographic Creative; 68, David Shale/naturepl.com; 69, David Shale/naturepl.com; 74, Bryce Groark; 77, Bryce Groark; 78-79, Mike Theiss/National Geographic Creative; 80-81, Flip Nicklin/Minden Pictures/National Geographic Creative; 82, Jo Mahy/age fotostock; 83, Alex Mustard/2020VISION/naturepl.com; 84-85, Brian Skerry/National Geographic Creative; 86-87, Jad Davenport/National Geographic Creative; 88-89, Scott Leslie/Minden Pictures/National Geographic Creative; 90-91, Brian Skerry/National Geographic Creative; 93, Alex Mustard/naturepl.com; 94-95, Alex Mustard/naturepl.com; 96, Alex Mustard/2020VISION/naturepl.com; 97, Alex Mustard/naturepl.com; 102, Kip Evans Photography; 105, Magnum Photos; 106-107, James L. Stanfield/National Geographic Creative; 108-109, Cesare Naldi; 110-111, Norbert Wu/Minden Pictures/National Geographic Creative; 112, Norbert Wu/Minden Pictures/National Geographic Creative; 113, Norbert Wu/Minden Pictures/National Geographic Creative; 114-115, Paul Sutherland/National Geographic Creative; 116-117, Thomas P. Peschak; 118-119, Laurent Ballesta; 120-121, Jad Davenport/National Geographic Creative; 123, Luis Miguel Cortes/National Geographic Your Shot; 124-125, Jad Davenport/National Geographic Creative; 126, David Shale/naturepl.com; 127, David Shale/naturepl.com; 132, Bryce Groark; 135, Brian Skerry/National Geographic Creative; 136-137, Andrew Watson/JAI/Corbis; 138-139, David Doubilet/National Geographic Creative; 140-141, Chris Newbert/Minden Pictures/National Geographic Creative; 142, Fred Bavendam/Minden Pictures/National Geographic Creative; 143, Melissa Fiene/National Geographic Your Shot; 144-145, David Doubilet/National Geographic Creative; 146, David Doubilet/National Geographic Creative; 148-149, David Doubilet/National Geographic Creative; 150, Chris Newbert/Minden Pictures/National Geographic Creative; 151, Fred Bavendam/Minden Pictures/National Geographic Creative; 152-153, David Doubilet/National Geographic Creative; 154-155, David Doubilet; 156, Justin Marshall/Queensland Brain Institute/University of Queensland; 157, Justin Marshall/Queensland Brain Institute/University of Queensland; 162, Arjan Gupta; 165, Yva Momatiuk & John Eastcott/Minden Pictures/National Geographic Creative; 166-167, Ralph Lee Hopkins/National Geographic Creative; 168-169, Ralph Lee Hopkins/National Geographic Creative; 170-171, Paul Nicklen/National Geographic Creative; 172-173, Paul Nicklen/National Geographic Creative; 174, Flip Nicklin; 175, Norbert Wu/Minden Pictures/National Geographic Creative; 176-177, Todd Gipstein/National Geographic Creative; 179, Flip Nicklin/National Geographic Creative; 180-181, Paul Nicklen/National Geographic Creative; 182-183, Brian Skerry/National Geographic Creative; 184-185, Bill Curtsinger; 186, Solvin Zankl/GEOMAR/naturepl.com; 187, Bill Curtsinger/National Geographic Creative; 192, Kip Evans; 195, Tui De Roy/Minden Pictures/National Geographic Creative; 196-197, George Steinmetz/National Geographic Creative; 198-199, Tim Laman/National Geographic Creative; 200-201, Christian Vizl; 202-203, David Doubilet; 204, Birgitte Wilms/Minden Pictures/National Geographic Creative; 205, Birgitte Wilms/Minden Pictures/National Geographic Creative; 206-207, Fred Bavendam/Minden Pictures/National Geographic Creative; 208-209, Norbert Wu/Minden Pictures/National Geographic Creative; 210, David Doubilet/National Geographic Creative; 211, Bill Curtsinger; 218, U.S. Fish and Wildlife Service/Ann Bell.

BLUE HOPE
Sylvia A. Earle

Published by the National Geographic Society
Gary E. Knell, *President and Chief Executive Officer*
John M. Fahey, *Chairman of the Board*
Declan Moore, *Executive Vice President; President, Publishing and Travel*
Melina Gerosa Bellows, *Executive Vice President; Chief Creative Officer, Books, Kids, and Family*

Prepared by the Book Division
Hector Sierra, *Senior Vice President and General Manager*
Janet Goldstein, *Senior Vice President and Editorial Director*
Jonathan Halling, *Creative Director*
Marianne Koszorus, *Design Director*
R. Gary Colbert, *Production Director*
Jennifer A. Thornton, *Director of Managing Editorial*
Susan S. Blair, *Director of Photography*
Meredith C. Wilcox, *Director, Administration and Rights Clearance*

Staff for This Book
Barbara Payne, *Editor*
Sanáa Akkach, *Art Director*
Matt Propert, *Illustrations Editor*
Carl Mehler, *Director of Maps*
Marshall Kiker, *Associate Managing Editor*
Judith Klein, *Production Editor*
Galen Young, *Rights Clearance Specialist*
Katie Olsen, *Production Design Assistant*
Michelle Cassidy, *Editorial Assistant*

Production Services
Phillip L. Schlosser, *Senior Vice President*
Chris Brown, *Vice President, NG Book Manufacturing*
Nicole Elliott, *Director of Production*
George Bounelis, *Senior Production Manager*
Rachel Faulise, *Manager*
Robert L. Barr, *Manager*

The National Geographic Society is one of the world's largest non-profit scientific and educational organizations. Founded in 1888 to "increase and diffuse geographic knowledge," the Society's mission is to inspire people to care about the planet. It reaches more than 400 million people worldwide each month through its official journal, *National Geographic,* and other magazines; National Geographic Channel; television documentaries; music; radio; films; books; DVDs; maps; exhibitions; live events; school publishing programs; interactive media; and merchandise. National Geographic has funded more than 10,000 scientific research, conservation and exploration projects and supports an education program promoting geographic literacy. For more information, visit www.national geographic.com.

For more information, please call 1-800-NGS LINE (647-5463) or write to the following address:

National Geographic Society
1145 17th Street N.W.
Washington, D.C. 20036-4688 U.S.A.

For information about special discounts for bulk purchases, please contact National Geographic Books Special Sales: ngspecsales@ngs.org

For rights or permissions inquiries, please contact National Geographic Books Subsidiary Rights: ngbookrights@ngs.org

Library of Congress Cataloging-in-Publication Data

Earle, Sylvia A., 1935-
  Blue hope : exploring and caring for earth's magnificent ocean / Sylvia A. Earle, founder of Mission Blue and National Geographic Explorer-in-Residence.
    pages cm
  ISBN 978-1-4262-1395-3 (hardcover : alk. paper)
  1. Marine ecology. 2. Deep-sea ecology. I. Title. II. Title: Exploring and caring for earth's magnificent ocean.
  QH541.5.S3E18 2014
  577.7--dc23

                              2014018878

Printed in the United States of America

14/CK-CML/1

LA MER
BLUE HEART

Partnering with the National Geographic Society for a third consecutive year
and inspired by "living legend" Dr. Sylvia Earle,
La Mer has embarked on a year-round celebration of the
world's oceans. Our Blue Heart campaign honors the work of emerging
explorers whose work embodies the beauty and power of the ocean.

Join us in raising awareness of the importance of healthy oceans
worldwide for ourselves and generations to come.

LaMer.com/WOD | @LaMerOfficial

LA MER